AF822806

Automated Process Planning for Additive Manufacturing with Directed Energy Deposition

Vom Promotionsausschuss der
Technischen Universität Hamburg

zur Erlangung des akademischen Grades
Doktor-Ingenieur (Dr.-Ing.)

genehmigte Dissertation (Monografie)

von

Vishnuu Jothi Prakash

aus
Pollachi

2025

1. Gutachter: Prof. Dr.-Ing. Claus Emmelmann
2. Gutachter: Prof. Dr.-Ing. Wolfgang Hintze

Tag der mündlichen Prüfung: 15. April 2024

Light Engineering für die Praxis

Reihe herausgegeben von

Claus Emmelmann, Hamburg, Germany

Technologie- und Wissenstransfer für die photonische Industrie ist der Inhalt dieser Buchreihe. Der Herausgeber leitet das Institut für Laser- und Anlagensystemtechnik (iLAS) an der Technischen Universität Hamburg (TUHH). Die Inhalte eröffnen den Lesern in der Forschung und in Unternehmen die Möglichkeit, innovative Produkte und Prozesse zu erkennen und so ihre Wettbewerbsfähigkeit nachhaltig zu stärken. Die Kenntnisse dienen der Weiterbildung von Ingenieuren und Multiplikatoren für die Produktentwicklung sowie die Produktions- und Lasertechnik, sie beinhalten die Entwicklung lasergestützter Produktionstechnologien und der Qualitätssicherung von Laserprozessen und Anlagen sowie Anleitungen für Beratungs- und Ausbildungsdienstleistungen für die Industrie.

Vishnuu Jothi Prakash

Automated Process Planning for Additive Manufacturing with Directed Energy Deposition

Springer

Vishnuu Jothi Prakash
Fraunhofer Research Institution for
Additive Manufacturing Technologies IAPT
Hamburg, Germany

ISSN 2522-8447 ISSN 2522-8455 (electronic)
Light Engineering für die Praxis
ISBN 978-3-031-87665-3 ISBN 978-3-031-87666-0 (eBook)
https://doi.org/10.1007/978-3-031-87666-0

This Springer imprint is published by the registered company Springer Nature Switzerland AG
The registered company address is: Gewerbestrasse 11, 6330 Cham, Switzerland

Abstract

Directed energy deposition (DED) technology is a resource-efficient tool-less manufacturing process with its characteristic high deposition rate and larger build volumes in comparison to its powder-bed counter-part. Combining the flexibility of a 6-axis industrial robot, complex 3-dimensional geometries can be realized with DED technology. This, however, poses a challenge in the process planning steps, namely data preparation step and build strategy determination due to insufficient software tools and lack of design standards. Additionally, the process planning steps for DED processes are part- and process-specific that have a direct influence on the quality and cost of the additively manufactured component.

In this work, three component clusters, namely multi-axis, freeform and thin-walled components were chosen not just for its relevance in the industry but also due to the prevailing tangible research gaps in the process planning of these component clusters. The industry de-facto STL file format-based data preparation has proven sufficient for powder-bed processes that are essentially two-dimensional. Additive manufacturing with multi-axis robot is rather three-dimensional requiring new slicing and toolpath planning framework and techniques. This thesis therefore begins with the development of a STEP file format-based slicing framework with which several component-specific toolpath generation algorithms have been developed, thereby offering great flexibility and precision in the developed toolpaths.

Followingly, thin-wall components that are important part families in aerospace sector have been investigated further. In addition to the development of novel toolpath planning algorithms, several process parameters and toolpath strategies have been optimized for geometrical accuracy and defect avoidance during deposition process and for the subsequent machining process.

Industrial demonstrators have been built to test and optimize the several algorithms developed within the framework of this thesis using a powder-based laser metal deposition system. This thesis aims to close some of the research gaps identified in the manufacturing planning phase of the DED processes, thereby enabling wide-spread adoption and industrialization of the resource-efficient manufacturing process.

Table of Contents

List of Figures

List of Tables

List of Symbols

Symbol	Description	Unit
α	Overhang angle	**deg**
μ	Viscosity	$\mathbf{kg/m.s}$
ρ	Density	$\mathbf{kg/m^3}$
σ	Surface tension	$\mathbf{N/m}$
σ_c	Surface tension at the center of the melt pool	$\mathbf{N/m}$
$\Delta\sigma$	Change in surface tension	$\mathbf{N/m}$
φ	Porosity	-
ξ, η, u, v	Parametric domain variables	-
A	Cross-sectional area	$\mathbf{mm^2}$
A_l	Cross-sectional area of the lower plane	$\mathbf{mm^2}$
A_o	Overhang area parameter	-
A_u	Cross-sectional area of the upper plane	$\mathbf{mm^2}$
b	Weld bead width	**mm**
C	Centroid	-
CA	Centroidal axis	-
C_p	Specific heat capacity	$\mathbf{J/kg.K}$
$C(\xi)$	Polynomial curve in the parametric domain ξ	-
d	Degree of a B-spline curve	-
d_l	Layer dilution	-
d_{spot}	Laser spot diameter	**mm**

d_w	Weld bead depth	mm
E	Energy Intensity	W/mm^2
$\dot{E}$	Energy per unit length	kJ/m
f	Aspect ratio	-
g	Gravitational field	m/s^2
h	Weld bead height	mm
h_5	Height after 5 layers	mm
k	Order of a B-spline curve	-
K	Thermal conductivity	W/mK
L	Link of a centroidal axis	-
$\dot{m}$	Powder mass flow rate	g/min
O	Origin	-
p	Pressure	N/m^2
P_l	Laser power	W
P_n	Evaluation point	-
$\vec{r}$	Normal vector	-
Q	Power generation per unit volume of the substrate	W/m^3
r	Radius of a circle	mm
$S\ (\xi, \eta)$	Surface defined in the parametric domain ξ and η	-
t	Time	s
T	Temperature	°C
U	Process velocity	m/s

List of Abbreviations

Abbreviation	Definition
2D	2-Dimensional
3D	3-Dimensional
AM	Additive manufacturing
AMS	Aerospace Material Specifications
API	Application Programming Interface
ASCII	American Standard Code for Information Interchange
ASTM	American Society for Testing and Materials
BREP	Boundary representation
CAD	Computer-Aided Design
CAE	Computer-Aided Engineering
CAM	Computer-Aided Manufacturing
CGAL	Computational Geometry Algorithm
CMM	Coordinate Measuring Machine
DED	Directed Energy Deposition
DIN	Deutsches Institut fuer Normung
DoE	Design of Experiments
DVS	Deutscher Verbund fuer Schweissen und verwandete Verfahren e. V.
EBAM	Electron Beam Additive Manufacturing
FAA	Federal Aviation Administration
GPU	Graphical Processing Unit

GUI	Graphical User Interface
IR	Infrared
ISO	International Organization for Standardization
LEAP	Leading Edge Aviation Propulsion
LMD	Laser Metal Deposition
MAT	Medial Axis Transformation
NURBS	Non-Uniform Rational B-Spline
NSGA	Non-dominated Sorting Genetic Algorithm
OCCT	Open Cascade Technology
OCT	Optical Coherence Tomography
PBF	Powder Bed Fusion
ROS	Robot Coordinate System
RPD	Rapid Plasma Deposition
RSM	Response Surface Methodology
TCS	Tool Coordinate System
TRL	Technology Readiness Level
STEP	Standard for the Exchange of Product Data
STL	Stereolithography / Standard Tessellation Language
WAAM	Wire Arc Additive Manufacturing
WCS	Workpiece Coordinate System

1. Introduction

Light-weight design, emission-free flight, environmental cleanliness and sustainable manufacturing are some of the current key research areas of the European aviation sector. Since 40 - 50% of the operational cost of a typical airliner is accounted by fuel, aviation industry is steered towards realization of lighter and more fuel-efficient aircrafts, for which these key research areas play a crucial role [1, 2]. Furthermore, the rising environmental awareness in developed nations steer the ministry and governing authorities to formulate the aforementioned research areas. Subsequently, new materials, smart systems, resource-efficient manufacturing technologies, alternate fuels and so on are being researched upon. Additive manufacturing (AM), also known as 3D-printing, which has been hailed as the 'third industrial revolution' has been an enabler to manufacture topologically optimized light-weight bionic structures for the aerospace and aviation industries [3–5]. By leveraging the advantages of this tool-less technology, metal AM processes not only enable resource-efficient manufacturing of aircraft structural and functional components, but can also make the existing production routes lean by eliminating some of the manufacturing processes and assembly steps [6–8].

Whilst many new metal additive manufacturing processes are being patented and developed in the recent years, powder bed fusion (PBF) and directed energy deposition (DED) processes are the best-known metal AM processes that have been already adopted by industries. Powder bed fusion processes uses either a laser or an electron beam source to selectively melt and fuse atomized metal powders spread evenly in an enclosed process chamber in a layer-by-layer manner [9]. Engineers have leveraged its design freedom, achievable surface and mechanical characteristics and functional integration to manufacture patient-specific implants, monobloc brake calipers, tools and dies with conformal cooling channels, etc. In combination with higher technological readiness level (TRL-8), a wide-spread industrialization of the technology can be witnessed [10]. Some of the notable industrial applications in aerospace sector includes the very first Leading Edge Aviation Propulsion (LEAP) engine fuel nozzle from GE Aviation to the very recent main oxidizer valve body for Merlin 1D Engine by SpaceX. A rising number of applications for serial production with an increasing number of machine tool manufacturers has been witnessed in the recent years. However, the technology is limited by its build volume, scalability and deposition rates.

With deposition rates as high as 2000 cu.cm per hour and build volumes exceeding 1 cu.m, directed energy deposition technology has gained focus and importance on the other end of the application spectrum [11, 12]. With the growing number of industrial-grade DED machines in the recent years, several applications have been witnessed in sectors like construction, tools and dies, transport and energy [13]. Notable applications have been witnessed in aerospace sector with the recent development of rocket hull by Relativity Space. Large volume components that otherwise require longer lead times through conventional manufacturing routes can be manufactured through the tool-less DED processes. In combination with the resource-friendliness, high material utilization and the multi-directional building flexibility of the process, sustainable parts with shorter lead time can be produced [14, 15].

V. J. Prakash, *Automated Process Planning for Additive Manufacturing with Directed Energy Deposition*, Light Engineering für die Praxis,
https://doi.org/10.1007/978-3-031-87666-0_1

1.1. Motivation

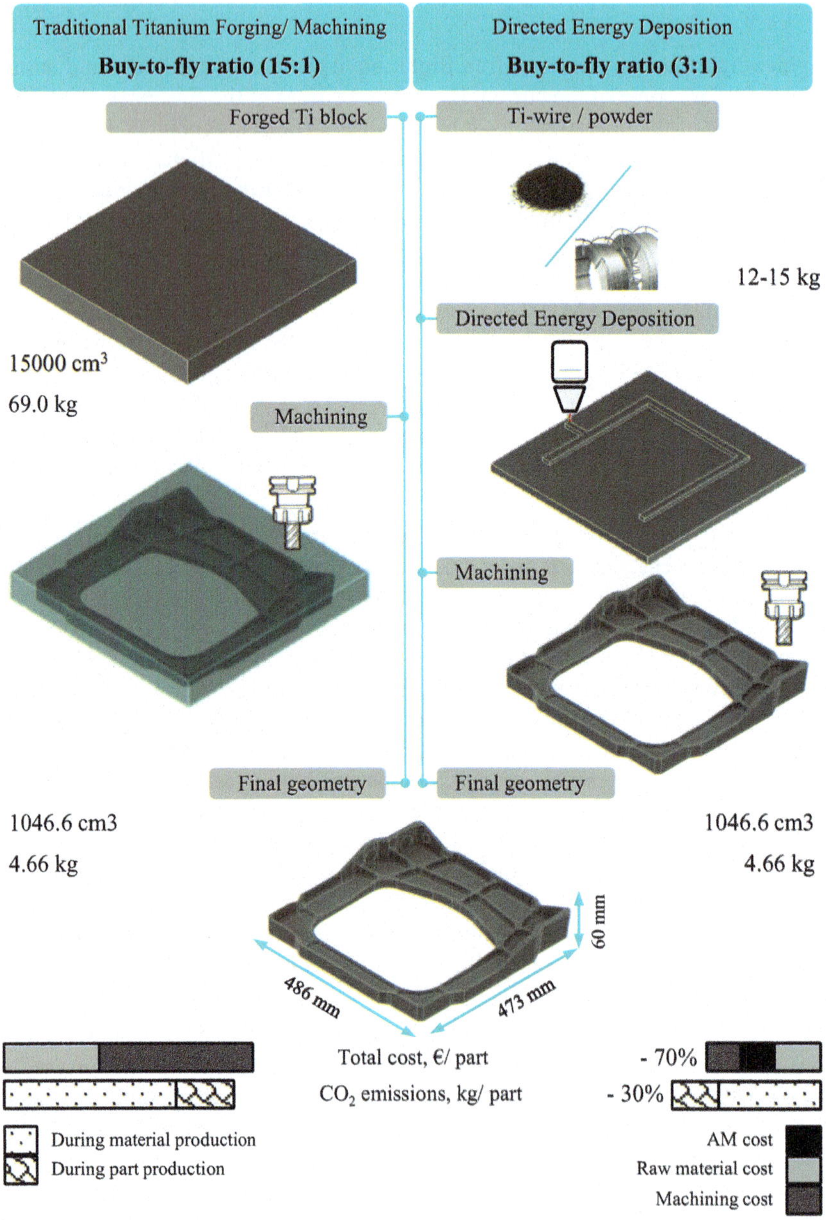

Figure 1.1 Advantages of directed energy deposition technology over conventional manufacturing processes with an aerospace demonstrator component

Aviation and aerospace sector have benefited from the advantages of the DED technology [2]. In particular, thin-wall components, which are characterized by large base surface area with rib-like strengthening structures are the most suitable part family for fabrication with this technology [16]. These thin-wall components are conventionally manufactured from wrought billets or forged pre-cursors followed by machining process. The typical buy-to-fly ratio in such cases is 20:1. Alternatively, fabrication of near-net shape pre-cursors through additive manufacturing followed by final machining reduces the buy-to-fly ratio to 3:1 or 2:1 (see Figure 1.1). The stated buy-to-fly ratio is based on the aerospace demonstrator component shown in the figure. Whereas, the total cost and the CO_2 emission comparison is based on an external study [17, 18].

Despite the promising economic advantages, a wide-spread industrialization is not yet seen in comparison to powder bed AM processes. Major hindrance in the implementation of this technology lies in the absence of reproducible process strategies [19]. Fabrication with high deposition rates and high energy inputs in combination with rapid cooling of the AM process leads to formation of residual stresses. These manifest into distortion, delamination and hot cracking, during the deposition process or in the downstream post-processing process [20]. Process and build strategies specific to process type (material form and energy source), machine kinematic, material, process parameter set, deposition rate and most importantly build geometry have proven to provide better results. Lack of this particular know-how in the existing CAD/CAM software packages is a major short-fall. Furthermore, there exists no design guidelines or methodologies that help develop process- and shape-dependent build strategies. Build strategies include identification of primary process parameters and optimization of toolpath. Though several applications of this technology had been witnessed in aerospace sector [21], it was not until 2017 when a structural part manufactured through DED process by Norsk Titanium was approved for serial production by the Federal Aviation Administration (FAA) for aviation purpose [22]. This research work aims to address the above-mentioned hindrances and to pave way for wide-spread industrialization of the DED technology.

1.2. Structure of the Thesis

The thesis begins with the current state-of-the-art analysis of several DED processes in the research domain and the achieved technology readiness level (TRL) in aerospace and aviation sector. Additionally, the availability of commercial and open-source CAD/CAM software packages and process planning platforms and their importance in industrialization of this technology are looked upon. Based on the literature review and industrial needs, key research areas have been identified and categorized under cost, quality and time, and are elaborated in chapter 3.

The research questions were organized and formulated into three goals that are then solved methodically. A systematic approach to achieve the research goals has been presented in chapter 4. Chapter 5 includes all the slicing and toolpath planning algorithms developed for identified component clusters and their implementations. The algorithms are developed to use STEP-file format instead of the industry de-facto STL-file format. To eliminate the resource-intensive trial-and-error approach for lot size 1 manufacturing, a methodology to systematically identify the process parameters including the slicing and hatch parameters has been developed and presented in chapter 6. The correlation between the primary process parameters with secondary factors (like component geometry, toolpath strategy, etc.) has been experimentally determined through fabrication of test geometries and demonstrator geometries followed by geometrical, metallurgical and mechanical characterization (section 6.2).

The thesis ends with concluding remarks and outlook on extendibility and transferability of this thesis to other metal AM technologies like powder-bed processes (see chapter 7). Figure 1.2 shows a schematic overview of the structure of the thesis.

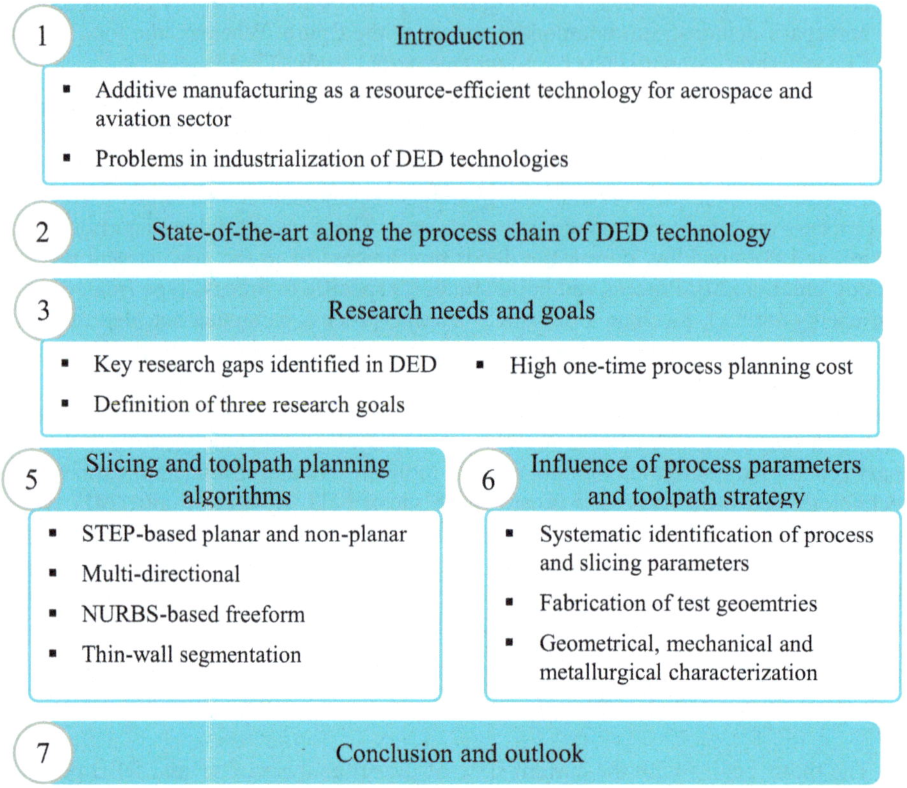

Figure 1.2 Overview and structure of the thesis

2. Literature Review

This chapter briefly describes additive manufacturing technology and its classification. The following section outlines the process mechanics, terminologies, classification and industrial use-cases of directed energy deposition technology. A generalized AM process chain is presented followed by an extensive literature review on the existing toolpath generation, planning and optimization research work specific to DED processes. This is complemented with an overview of the commercially available CAD/CAM software packages for process planning of multi-axis DED processes and their limitations.

2.1. Additive Manufacturing

ASTM 52900 defines additive manufacturing as the "process of joining materials to make parts from 3D model data, usually layer upon layer, as opposed to subtractive manufacturing and formative manufacturing methodologies" [23]. Fundamental principle of AM is the successive addition of material for shaping of a three-dimensional object. Depending upon how the layer-wise build-up takes place and similarities in material processing, AM processes can be categorized into seven process categories as defined by the ASTM F42 committee: binder jetting, directed energy deposition, material extrusion, material jetting, powder bed fusion, sheet lamination and vat photopolymerization. These processes can additionally be clustered or ranked based on one or many of the criteria mentioned below:

1. **deposition rate**, defines the additive volume deposited over time. Typically measured in cu. cm/hr or sometimes kg/hr for high deposition rate processes
2. **machine hour-rate**, defined by the machine cost and other parameters like depreciation cost, fixed and variable overheads, etc.
3. **build resolution**, defines the smallest possible feature that can be manufactured. It is defined by mm or µm.
4. **build volume**, defines the dimensions of the build chamber or the maximum possible build dimensions.

In comparison to other shaping processes like formative and subtractive, additive manufacturing processes has the following advantages:

Design freedom: Layer-wise fabrication allows designers and engineers to use this modern manufacturing approach to realize lightweight components and bio-inspired structures. By harvesting the design freedom of AM together with biomimetics, novel load optimized structural components and artifacts have been made possible.

Functional integration: In addition to design flexibility, integration of features and complex structures like conformal cooling channels in molds and dies increases product efficiency. Furthermore, part consolidation minimizes the total number of single parts, its manufacturing process steps and could also eliminate down-stream assembly.

Tool-less manufacturing: AM is a tool-less manufacturing technique requiring no additional dies or molds, thereby reducing the cost and time required to manufacture them.

V. J. Prakash, *Automated Process Planning for Additive Manufacturing with Directed Energy Deposition*, Light Engineering für die Praxis,
https://doi.org/10.1007/978-3-031-87666-0_2

This makes the process more suitable for prototype development and manufacturing of individualized components with lot size 1.

Resource-efficient manufacturing: Alloys of titanium, nickel and other light-weight, high-strength metals when manufactured conventionally via subtractive processes leave material wastes that are not re-usable. Usage of AM processes in combination with light-weight design methodology and powder re-usability enables resource-efficient manufacturing.

2.2. Directed Energy Deposition

Directed energy deposition is defined as 'an additive manufacturing process in which focused thermal energy is used to fuse materials by melting as they are being deposited' according to ASTM F3187 [24]. DED is essentially a welding process used for generating 3D metallic components from CAD data by layered deposition. A deposition head housing the necessary optics, electronics and nozzles, traverses a pre-programmed vector path of one layer and then translates to the next layer and continues the building process on the previous layer, thereby generating a 3D component [25–29]. The deposition process starts with a substrate, usually a build plate of the same material or a pre-fabricated component in case of hybrid manufacturing. During the deposition process, an energy source (laser, arc or electron beam) creates a melt pool in the substrate (or pre-built layer) into which material is injected through nozzles and the melt pool then undergoes a sequence of melting and solidification cycle. Figure 2.1 shows a simplified schematic representation of a directed energy deposition process. The figure shows the multi-directional, support-free deposition of material over a substrate clamped on to a turn-tilt table.

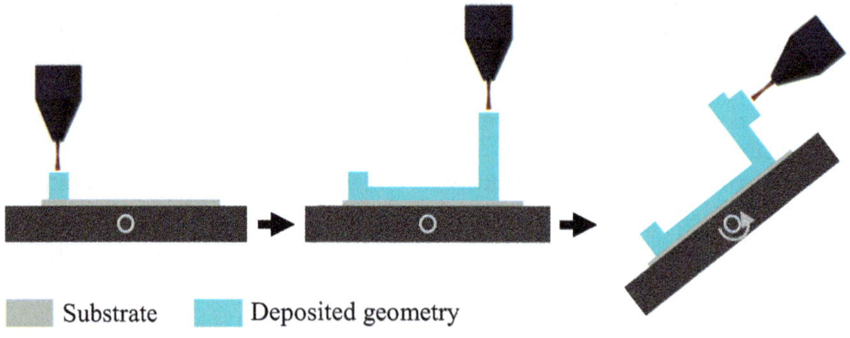

Figure 2.1 Schematic representation of a multi-directional, support-free directed energy deposition process

DED technology can be classified based on the energy source used and the form of material employed. Figure 2.2 gives an overview of the different DED processes along with the machine tool manufacturers. Based on the total number of industry-grade machines, the most common process amongst them is the laser metal deposition (LMD) with powder or wire, owing to its repair applications. Other processes include wire arc additive manufacturing (WAAM), electron beam additive manufacturing (EBAM) and plasma-arc based deposition, which are primarily used for additive manufacturing of 3D-structures. Recent additions to the list are cold spray, ultrasonic motion, joule printing and friction-stir welding based deposition. Deposition rates of the DED processes range between 100 cu. cm/hr and 8000 cu. cm/hr and the build dimension ranges from 500 x 500 x 500 cu. mm to 7000 x 1200 x 1200 cu. mm. or even more in case of robotic systems mounted on a linear drive [30]. Figure 2.3 shows a relative comparison of PBF and DED technology in terms of build volume or deposition rate and the achievable build resolution or shape complexity.

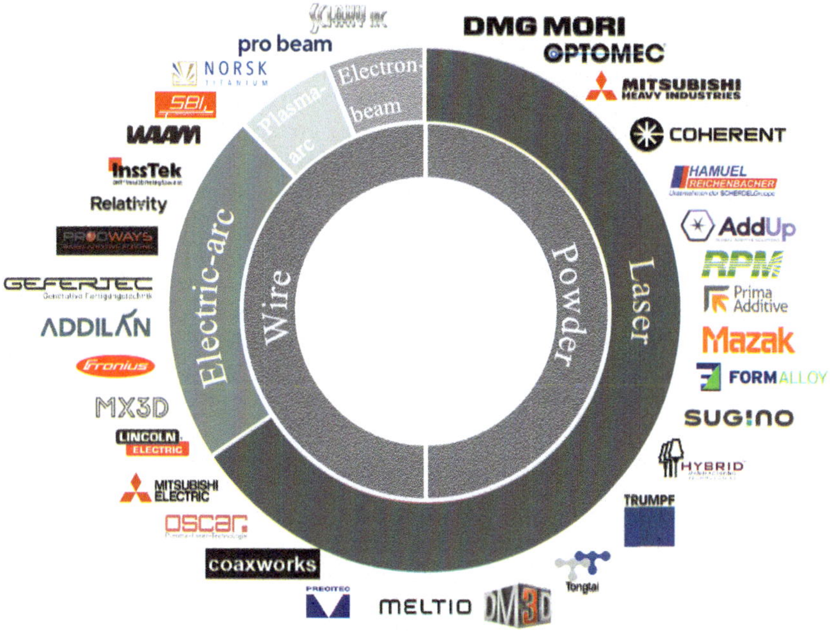

Figure 2.2 Classification of DED processes and an overview of machine manufacturers

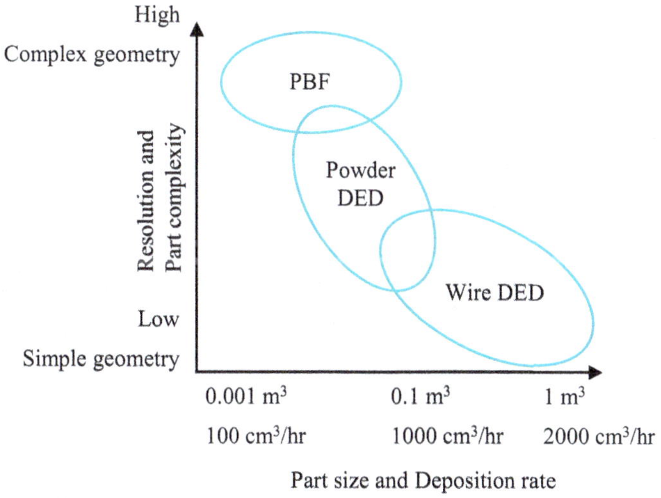

Figure 2.3 Part complexity vs. Deposition rate comparison between powder bed fusion process and directed energy deposition process (based on [31])

2.3. Laser Metal Deposition

In this research work, powder-based laser metal deposition has been used for implementation and investigation of several algorithms and the developed methodologies. Process mechanics of the LMD process has been briefly described in this section, whereas the system setup and the material used for additive manufacturing of test geometries and demonstrators are described in section 6.1. Laser metal deposition process is a multi-layer continuous welding process using laser as source to create a three-dimensional structure [21, 26]. The so-called deposition head is equipped with optical elements to collimate laser beam onto the substrate (or base plate) where atomized powder particles are injected coaxially through a nozzle system. The thermal energy creates a melt pool, which then rapidly solidifies. By traversing the deposition head in a pre-determined path in combination with a laser control system, three-dimensional structures can be built. The deposition head is typically mounted to a kinematic system, such as a 3-axis gantry or a 6-axis industrial robot. The substrate is a base plate of the same or an alloy of the same material that is to be deposited. In some cases, the substrate can be more complex than a plate, for example, a forged or a machined component, over which additional features can be deposited. Figure 2.4 shows the schematic representation of an LMD process with the associated coordinate systems, namely, workpiece coordinate system (WCS), robot coordinate system (ROS) and tool coordinate system (TCS).

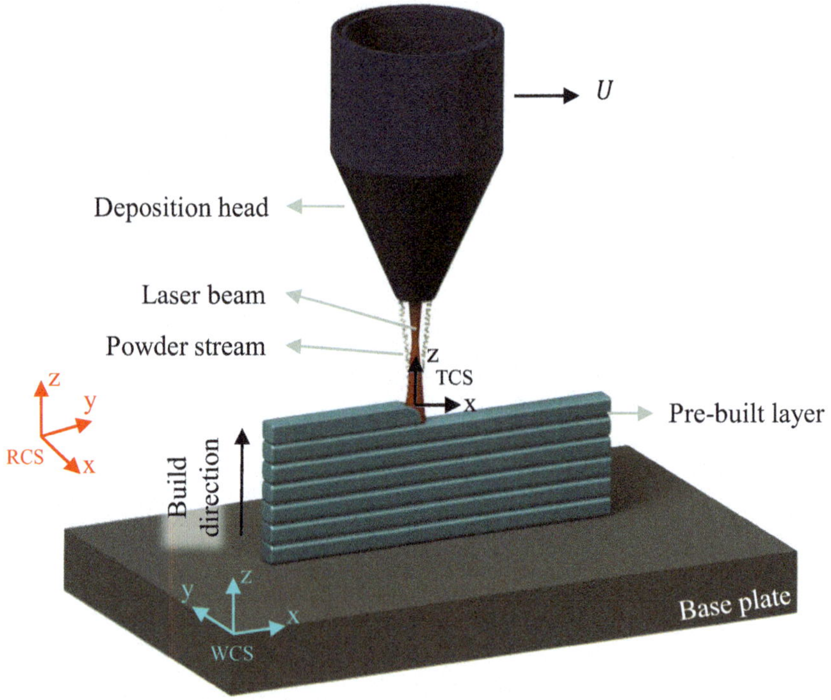

Figure 2.4 Schematic representation of a laser metal deposition process with the associated coordinate systems (recreated based on [14])

2.3.1. Process Modeling

Fabrication through laser metal deposition involves the following sequential physical phenomena: laser absorption, heat conduction, metal powder addition, melting and rapid solidification. Thermal energy generated when the laser hits the substrate are absorbed by the substrate to form the melt pool. The stream of atomized metal powder that is being sprayed absorbs a portion of this energy before it is added into the melt pool [32]. A heat conduction equation is used to model this phase of the process. The gaussian mode laser beam that is used during the deposition process creates maximum energy intensity at its center that decreases along the radial direction. This leads to a temperature gradient on the surface of the melt pool which further leads to surface tension (σ) gradient. This eventually generates a convective flow within the melt pool driving the molten metal outwards, i.e. from the central low surface tension region,σ_c to the edges with high surface tension, $\sigma_c + \Delta\sigma$. This surface tension driven convective flow is known as Marangoni convection (see Figure 2.5) [33]. This phase of the process is expressed using momentum, heat transfer and continuity equations. The transient temperature distribution at any arbitrary point and time, defined by T(x,y,z,t), is obtained from the three-dimensional heat conduction equation as shown in equations 2.1 and 2.2 [34].

$$\frac{\partial(\rho c_p T)}{\partial t} + \nabla.\left(\rho c_p U T\right) - \nabla.(K\nabla T) = Q \tag{2.1}$$

$$\frac{\partial(\rho T)}{\partial T} + (\rho U\nabla)U = \rho g - \nabla p + \mu\nabla.(\nabla U) \tag{2.2}$$

where,

Q, power generation per unit volume of the substrate $[W/m^3]$
K, thermal conductivity $[W/mK]$
C_p, specific heat capacity $[J/kg.K]$
ρ, density, $[kg/m^3]$
U, process velocity $[m/s]$
g, gravitational field $[m/s^2]$
μ, viscosity $[kg/s.m]$
p, pressure $[N/m^2]$

Thermo-mechanical models are used to simulate the layer-wise material deposition process in terms of temperature distribution, cooling rates and the thereby resulting deformation and residual stresses [35]. Several analytical and numerical approaches (like Finite Element Modeling) have been developed as a function of primary process parameters (like energy intensity and material feed rate) to simulate the thermal behavior of the process, and to predict the resulting microstructure that determines the material characteristics of the component [36–39]. Process modeling and simulation models help understand the process better and to optimize the process parameters that ensure process stability and to achieve the required static and dynamic mechanical characteristics [40, 41]. For example, the accumulated heat input over the build process and the reduction in heat conduction with the base plate as the deposition process moves far away from the base plate can be simulated in-prior and an optimized layer-wise adaptive energy input could be developed [42].

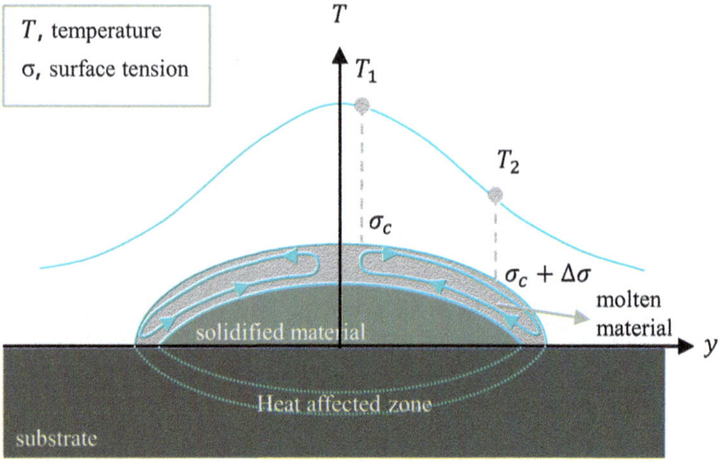

Figure 2.5 Schematic representation of the melt pool region and the Marangoni convection within the melt pool (recreated based on [43])

2.3.2. Process Control

Even though simulation models predict process characteristics and optimize process parameters to a certain extent, the exhaustive list of intrinsic and extrinsic process parameters and their interaction are not considered in simulation models, leading to process deviations [44]. One of the assumptions made in thermo-mechanical models is the pre-determined constant deposited layer geometry (eventually the weld bead dimensions like width and height). Changing thermo-mechanical boundary conditions during the deposition process results in changing deposited layer height and width over the build height and at certain geometrical features like corners where the actual deposition speed varies due to deceleration and acceleration of the kinematic system or at intersections due to material overlap. This results in deposition nozzle being either too close or too far from the last built layers, thereby deviating from the ideal *working distance* [45]. On the other hand, the CAD geometry is sliced with a constant slice thickness and so are the toolpaths generated with a constant pre-determined z-offset. Deviation of working distance from its ideal value leads to powder stream not converging at the desired location and a rather different energy distribution of the laser beam. This leads to change in material deposition volume and geometry that will eventually lead to process instability and build job termination. A schematic representation of the same is shown in Figure 2.6.

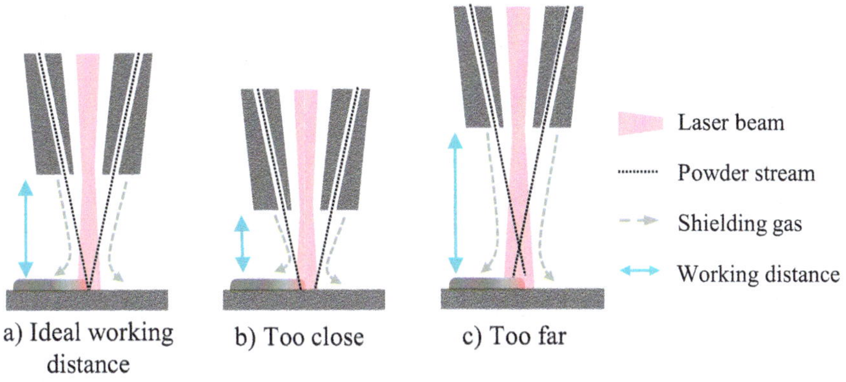

Figure 2.6 Schematic representation of the laser metal deposition process with working distance that is: a) ideal; b) too close and c) too far

Volumetric modeling approaches have been developed to predict the layer geometry using mass balance equations [46–48]. These models can predict the layer height and width, from which layer-wise variable process parameters (like slice thickness, laser power and deposition velocity) can be derived, which ensures a constant working distance. These models are however limited to simple geometries and are time-consuming and computationally intensive [49]. Additionally, environmental factors cause variation in the deposited layer geometry and melt pool dynamics [50, 51]. Therefore, in industrial scenario, one or several online process monitoring tools like laser triangulation sensor, optical coherence tomography sensor, thermal and image sensors are utilized [51–56]. These process monitoring systems have proven to be effective and could be further used for process control applications [56, 57].

2.4. DED Process Chain

A generic process chain of the DED process can be represented as shown below in Figure 2.7.

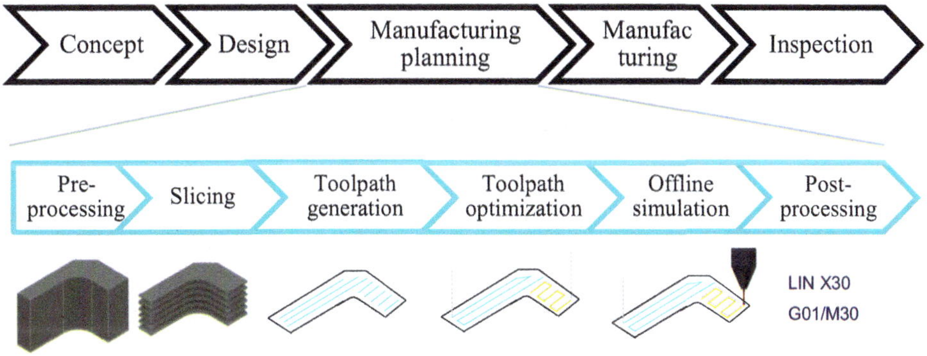

Figure 2.7 Process chain of a typical DED process highlighting the digital work-flow in the manufacturing planning phase

A detailed work flow from the concept phase to the inspection phase with a list of all the influencing process parameters and constraints to consider is represented schematically and explained briefly in section 4.1. This section of the chapter highlights, in particular, the work flow in the manufacturing planning phase (refer Figure 2.7). In this phase, the CAD model of the workpiece (also referred as component or geometry) to be additively manufactured is converted into a machine-readable file with the necessary toolpath and controller information. The workflow begins with the so called 'slicing' process, where the three-dimensional model is intersected with a cut-section plane that is translated along the build direction, resulting in a series of two-dimensional intersection geometries, known as 'slices'. The translation distance is equal to one-layer deposition thickness of the chosen AM process [49]. Followingly, for every slice, toolpaths are generated that the deposition head traverse during the process. These toolpaths are commonly known as 'hatch paths' in powder bed processes. Being a multi-directional, support-less deposition process coupled with a multi-axis kinematic system, DED technology enables fabrication of three-dimensional components. This advantage gives rise to the need of CAD model pre-processing steps, new slicing techniques, toolpath generation and optimization approaches very specific to the geometry.

In this research work, components or geometries are grouped into six different clusters sharing similar geometrical characteristics. Figure 2.8 shows an overview of the manufacturing planning phase specific for the six component clusters. Pre-processing of CAD model like feature recognition or sectioning is required for certain part geometries before slicing. This step is very specific for DED technology and is not necessary in powder bed fusion processes. Following the slicing and toolpath generation steps, the generated toolpaths are often optimized for one or many of the optimization functions, namely, cost, deposition time, uniform heat distribution, collision avoidance, etc.

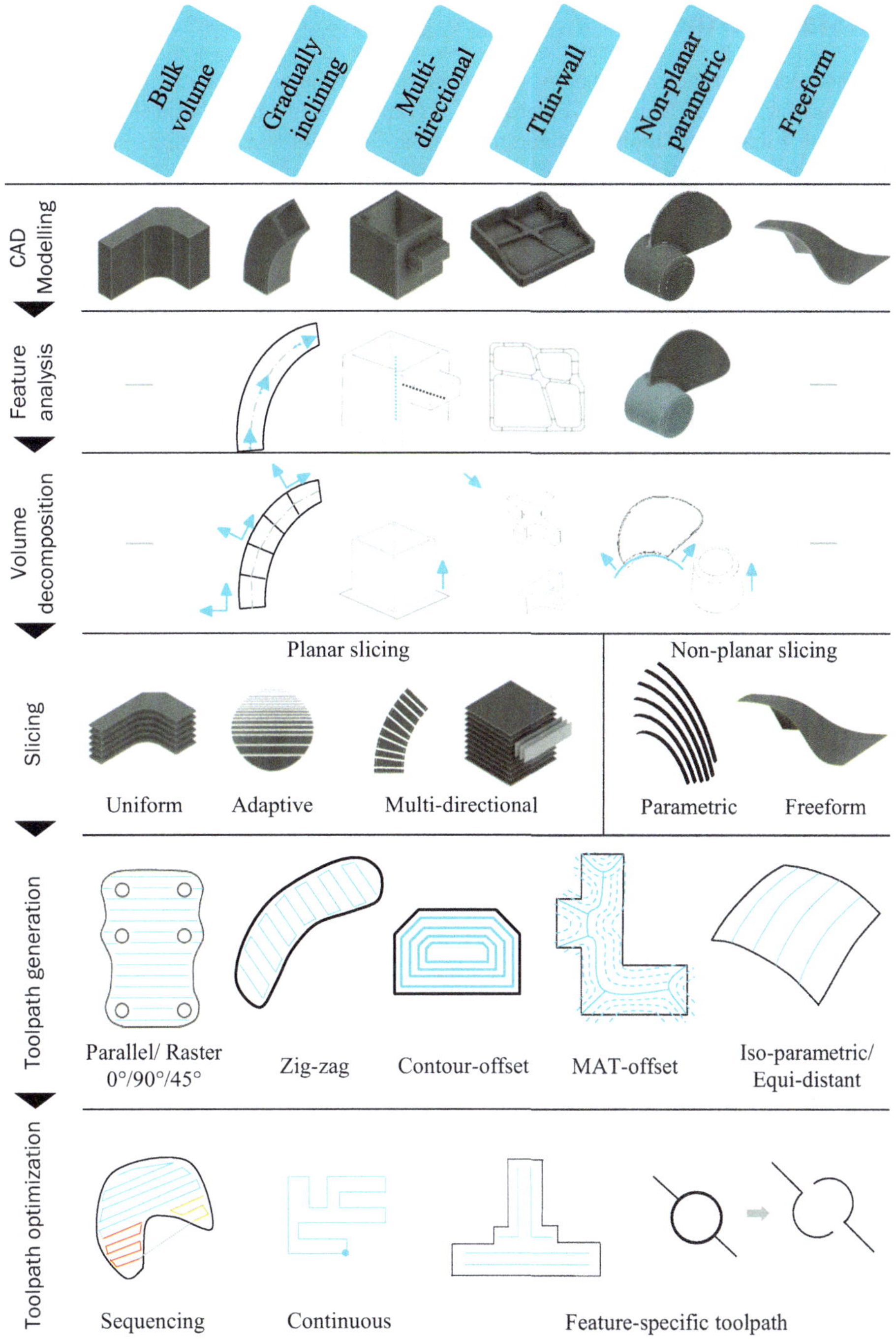

Figure 2.8 Overview of the toolpath generation and optimization phase for different component geometries and for different slicing approaches

2.5. State-of-the-art

This section of the chapter consolidates the state-of-the-art in the manufacturing planning phase within DED process chain, i.e. CAD model analysis, volume and feature segmentation, slicing and toolpath planning, both in the academia as well as commercially available software packages.

2.5.1. Research Activities

Research activities carried out in the academia is mainly focused on the development of various algorithms for CAD-data handling including slicing and toolpath preparation. The developed algorithms were implemented either using C++ or Matlab or Python. Most of the literature collected were focused on one particular DED process type and one specific component geometry.

Table 2.1 shows an overview of the collected literature classified along with a summary and identified research gaps in accordance to the process chain represented in Figure 2.8.

Table 2.1 Overview of the collected literature classified along the process chain along the identified research gaps

Classification		Ref.	File format	Description
Feature analysis and volume decomposition				
Planar	Multi-directional	[58]	STL	Volume segmentation of mesh-based geometries through silhouette edge projection and analysis of surface normal vectors
		[59, 60]	STL	Centroidal-axis based volume segmentation to enable multi-directional slicing
		[61–64]	STL	Volume segmentation based on cusp-height detection and volumetric difference between layers
		[65, 66]	STL	Curvature-based volume decomposition and depth-tree structure for regrouping and sequencing of geometries
		[67]	STL	Random cutting plane approach for feature analysis and volume segmentation
		[68]	STL	Volume decomposition by Reeb graph and curvature analysis; Optimized build direction search
		[69]	STL	Adaptive conformal trajectories with connectivity graph and attribute adjacency matrix
	Gradually inclined	[70]	STL	Multi-directional mesh segmentation for overhang features; Minimum overhang length as objective function for optimization of build direction for every segment
		[71]	STL	Non-planar 3D path with constant layer geometry for support-less cantilever tube deposition

	Thin-wall	[72]	STL	Medial axis-based feature analysis for automotive panel application
Non-planar	Parametric	[73, 74]	Native file - CATIA	STEP entity-based volume recognition; Developed as a CATIA plug-in; limited to two STEP entity features
Summary and research gap				There exist several STL-file format-based algorithms for volume segmentation, followed by multi-directional planar slicing. Of the several computational techniques used, centroidal-axis based segmentation is proven to be robust. There exist little or no research paper that focus on volume decomposition of thin-wall geometries and non-planar geometries. Furthermore, there exists no STEP-file format-based volume decomposition algorithm
Slicing				
Planar	Uniform	[64, 75–78]	STL	Slicing of mesh-based tessellated geometries with uniform slicing thickness
		-	STEP	-
	Adaptive	[79–84]	STL	Slicing of mesh-based tessellated geometries with adaptive slicing thickness to improve surface roughness by minimizing cusp height
		[85]	STL	Inclusion of normal vectors for 5-axis deposition
Non-planar	Cylindrically	[74, 63]	STL	Cylindrical non-planar surface slicing of mesh-based geometries for deposition of rotational components
	Spherically-	[86]	STL	Spherical surface-based slicing of mesh-based geometries
	Direct	[87–93]	STEP	Slicing of native file format within a CAD/CAM software tool like Powershape, Unigraphics and CATIA; Developed as a plug-in

Summary and research gap				STL-file format-based slicing is an industry standard that is commonly prevailing in both commercial software applications and in the academia. The first very algorithms were developed in the early 1990s and over time several optimization algorithms have been developed that increase the robustness and decreases the computational time. The capabilities have been extended to slice with a cylindrical or a spherical surface as well. Few STEP-file format-based slicing algorithms have been seen being implemented within an existing CAD/CAM software tool through its application programming interface (API). There exists no full-fledged STEP-file format based slicing library for non-planar geometries.
Toolpath generation				
Planar	Parallel/ Raster/	[94–98]	STL	Parallel-scanline based toolpath generation with implementation of zig or zig-zag strategy; Algorithms for region sequencing; Algorithm for efficiency improvement; GPU-accelerated algorithms
	Contour-offset	[99–101]	STL	Development of contour offsets from outside to inside
	Continu-ous/ Hil-	[102]	STL	Development of an infill pattern with a single continuous toolpath, thereby eliminating rapid motions between toolpath segments; Requirement of certain extrusion-based AM processes
	MAT-offset	[103–106]	STL	Offset curves using medial axis transformation information (MAT) from inside to outside
Non-planar	-	-	-	
Summary and research gap				Several toolpath (hatching) strategies have been developed essentially for powder-bed processes, which are also suitable for directed energy deposition of planar geometries. Additionally, medial-axis based toolpath generation algorithms have been developed for thin-wall geometries, in which case, the MAT of the entire geometry is

				considered. Feature detection is still a manual process that needs automation. There exist no academic research work or algorithm or library that enables generation of non-planar toolpaths from STEP-based file format. However, there exists non-planar toolpath generation tools for machining of freeform surfaces[107–114].
Toolpath optimization				
Planar	Region-sequencing	[115]	STL	Region sequencing of raster toolpaths; Generation of optimized connecting lines for zig-zag strategy; Contour-raster void optimization
	Thin-wall	[116]	STL	End lateral extension path strategy for intersection regions
		[117, 118]	STL	Extraction of medial axis information, followed by generation on wavy zig-zag path along medial axis to accommodate varying wall thickness
		[118]	STL	Manual segmentation of thin-wall geometry features followed by toolpath planning; Zig-zag and parallel toolpaths for segmented features
		[119]	STL	Continuous eulerian paths for thin-wall geometries to avoid voids
		[120]	STL	Toolpath approaches for multiple track intersections
		[121]	STL	Thin-wall toolpaths for blade geometries
		[122]	STL	Reinforcement learning and sequential least square programming optimization-based tool-path planning
		[123]	STL	Multi-criteria optimization for region sequencing and to optimize distance and time
	Time-optimal	[124]	STL	Time-optimized toolpath based on cost model
	Build up strategies	[125]	STL	Zig-zag toolpath strategy with variable track overlap distance, inter-layer wait time, improvement of shielding gas flow
	Mechanical-stress	[126]	STL	Alignment of extrusion direction with high stress tensors to reduce the effects of weak inter-layer bonds
		[127]	STL	A 3D printing tool-path generation strategy based on the partition of principal stress field for fused

	filament fabrication estimated through Finite Element Analysis
Summary and research gap	Optimization of toolpath has been one of the recent research areas in the field of DED technology along with the study of its effect on the build quality, cost and time. Several application-specific toolpath optimization strategies have been developed for thin-wall aerospace demonstrators.

2.5.2. Commercial Software Packages

Figure 2.9 Classification of commercially available software packages (or modules) for Directed Energy Deposition process

Throughout the AM process chain, several CAD/CAM software packages or tools are being used. A list of commercially available slicing and toolpath planning software packages for DED processes are shown in Figure 2.9. These software packages are classified under three categories, namely:

1. Process- and machine-specific software packages: Unavailability of commercial software packages in the early years led to the development of software tools by some of the DED machine tool manufacturers that are specific to the process and the machine tool developed.
2. DED technology-specific software packages: To address the inadequacies in the multi-axis deposition modules developed by the generic CAD/CAM software developers, few DED-technology specific software packages have been developed in the recent times.
3. Generic CAD/CAM software packages: With the increasing number of DED machine tool manufacturers in the recent years combined with the raising industrial use-cases, generic CAD/CAM software developers have developed multi-axis deposition modules/ add-ins to their product portfolio to address this market.

Table 2.2 lists the advantages as well as the limitations of the above-categorized commercial software packages.

Table 2.2 Advantages and limitations of the categorized commercially available software packages

Process- and machine-specific software packages	
+	Tailored to specific DED process, includes design rules
+	Consideration of machine-tool kinematics and directionality of process head (if applicable)
+	Includes manufacturer's process know-how
+	Includes material database and its corresponding process parameter sets
–	Usually developed by the machine manufacturer with limited software development capabilities
–	Comparatively low-level GUI with no much importance to user-friendliness
–	Limited or no design and modelling functionalities
–	Not easy to integrate into company's CAE infrastructure
–	Supports only STL-file exchange format
DED technology specific software packages	
+	Tailored for DED processes, includes technology-specific data pre- and post-processing functions and algorithms
+	Robot kinematics of major manufacturers included, thereby considering the limitations of machine-tool kinematics
+	Can be tailored for any DED process or kinematic setup
+	Also supports STEP-file exchange format
–	Not easy to integrate into company's CAE infrastructure
–	Includes limited process-specific know-how
–	Lacks material database
Generic CAD/CAM software packages	
+	Additional DED module to existing software; Integration into company's existing CAE eco-system is made easier
+	End-to-end digitalization of process chain possible
+	Robotics module already included
+	Supports native and STEP-file exchange format
+	Built upon powerful, stable, existing geometric modeling kernel with a well-known GUI framework
–	Not tailored to any specific DED process; Comparatively limited toolpath generation and optimization functionalities
–	Includes neither process know-how nor material database
–	Licenses are expensive

2.6. Exchange File Format

2.6.1. STL-file Format

Developed at 3D Systems in 1980s, this file format is the most widely used exchange file format in the AM industry till date [77]. The surface of a 3D CAD model is tessellated into tiny triangles called facets which are then stored in a file, either as ASCII format or binary format. Both these formats contain information of the coordinates of the three vertices $V(x_{1,2.3}, y_{1,2.3}, z_{1,2.3})$ and the unit normal vector $N(x_n, y_n, z_n)$ of the facets, pointing outwards in accordance to the 3D model. Figure 2.10 shows the CAD model of a cube and the tessellated geometry of the same. One of the facets is shown in the figure as well with its vertices and the normal vector according to the right-hand rule.

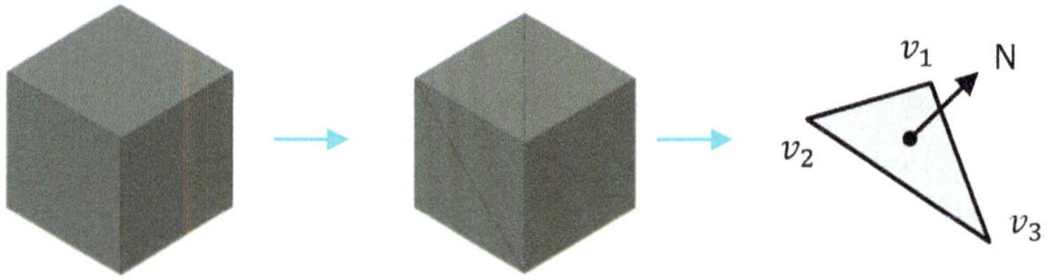

Figure 2.10 3D CAD model of a cube in its native file format and as a tessellated geometry. Schematic representation of a tessellated facet with its vertices and a normal vector

Slicing and toolpath extraction of stereolithographic file format (.stl) data has been an industry de-facto for decades. Single directional slicing strategy slices the tessellated file in the build direction, usually the Z-axis, with the algorithm introduced in the early 1990s [78, 77, 64]. The algorithm checks for intersections of the triangles with a slicing plane which translates along the build direction by the distance equal to the user-defined slicing thickness. The intersection points (connected to form a line) are then sorted from head-to-tail and connected to form a contour of the layer. This is followed by the generation of hatch lines (toolpaths) to fill the inner geometry. Both the tessellation of the CAD model and the uniform slicing procedure introduces inaccuracies in the printed geometry. The smaller the edge length of the facets, the higher is the geometrical accuracy [76, 89, 88].

Powder bed AM processes are technically layer-wise 2D printing process stacked in the 3rd dimension, thereby eliminating the need of normal vector information at any arbitrary location during the manufacturing process. The build direction remains constant through-out the process, as is restricted by the machine kinematics. On the contrary, most of the DED processes are essentially multi-directional deposition systems capable of fabricating freeform components without the requirement of support structures. This, however, requires additional information, like normal vector, during the fabrication process as well.

2.6.2. STEP-file Format

Unlike STL-file format which contains surface information as triangulated meshes, Standard for the Exchange of Product Data or STEP (ISO 10303) exchange file format entails geometry information like points, vectors, directions, curves and surfaces as well as topological entities like vertices, edges, loops, faces, shells, volume and solids with high geometrical accuracy. The geometrical entities are individually defined, positioned, oriented and are also connected through an associated topological graph [89, 88]. Such a neutral file format provides a complete and accurate description of the CAD model, including the analytical and mathematical description of the features. Figure 2.11 shows the STEP model of a cube stored in the form of Boundary Representation modeling. Instead of tessellating the model into smaller facets, the solid volume is composed of six surfaces or faces, wherein each face is composed of four edges with two vertices forming an edge.

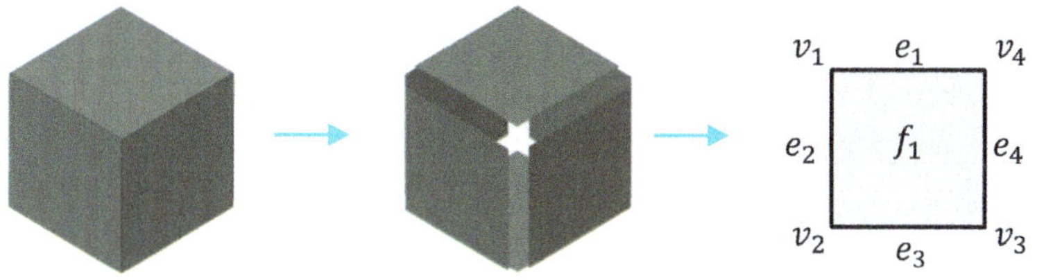

Figure 2.11 3D CAD model of a cube in its native file format and its representation as a STEP geometry, with its geometrical and topological elements

To work with native file formats or neutral file formats like STEP, a so-called geometric modeling kernel is essential. Such a geometric kernel is a software component or a library that encompasses the necessary classes for handling basic data structures, modeling data and basic modeling algorithms with which a computer-aided design package can be developed. Several geometric modeling kernels are available in the market, amongst which, Parasolid by Siemens, ACIS by Spatial Corporation and ShapeManager by Autodesk are the widely used and adopted kernels. In this work, the only available open source geometric modeling kernel namely Open Cascade Technology (OCCT) has been used for reading and processing STEP-file format-based CAD models. A brief overview of the OCCT software library and the algorithm developed for slicing of STEP-file format-based CAD model is described in section 5.1.

3. Research Needs and Goals

Despite several years of research work carried out in the field of DED technology, there has been no wide-spread industrialization and application in aviation and aerospace sectors. Whereas, powder bed processes have achieved higher technology readiness level (TRL) and are integrated into manufacturing process chain by several Original Equipment Manufacturers and Tier 1 companies [18, 10].

Notable DED applications in the aerospace sector include the fabrication of outer hull of the rocket by Relativity Space using wire-arc additive manufacturing (WAAM) process, Sciaky's use of their proprietary electron beam additive manufacturing (EBAM®) process in vacuum atmosphere for fabrication of satellite tanks and MT Aerospace's application of powder-based laser metal deposition system from AddUp for satellite components [10]. As of today, Norsk titanium, with its headquarters in Norway, is the only company to qualify their proprietary wire-fed rapid plasma deposition (RPD®) process for titanium in the aviation industry. The plasma deposition process developed by them takes place in an inert gas atmosphere equipped with a closed-loop sensor and control system. In 2015, the company managed to reach TRL 8 with their process, where as in 2017, their process has been certified by the Federal Aviation Administration (FAA) for additive manufacturing of aviation components for Boeing Dreamliner 787 [22].

3.1. Key Research Areas

Four key research areas are identified that enable industrialization of DED technologies especially in aerospace and aviation sector, and they are enumerated below:

1. **Process understanding:** Development of a better understanding of the thermo-mechanical behavior of the deposition process and the influence of process parameters on the resulting mechanical properties of the fabricated geometry. By far, this is the most researched area in directed energy deposition technology. However, the correlations between the process parameters and toolpath strategy for high deposition rate and their effect on geometrical accuracy, thermal and microstructural behavior in an industrial use-case component is not yet fully studied [13].
2. **Lack of design guidelines and standards:** Even though the process can be dated back to two decades, there are no sufficient design guidelines, standards and norms that list quality parameters like allowable pore size or defect size, grain structure and minimum achievable static and dynamic mechanical properties of the fabricated components.
3. **CAX software framework:** It was not until early 2020s when commercial and mainstream CAD/CAM software packages included a module for toolpath generation and planning for multi-axis deposition processes. Lack of process and system independent CAD/CAM software packages hindered automation and scalability of the process in an industrial setup. These packages, however, still lack process-specific know-how and component-specific design automation algorithms. For research purposes, these software packages lack the flexibility to tweak individual toolpaths critical for build quality.

V. J. Prakash, *Automated Process Planning for Additive Manufacturing with Directed Energy Deposition*, Light Engineering für die Praxis,
https://doi.org/10.1007/978-3-031-87666-0_3

4. **Quality control and assurance:** Reliable sensors and closed loop control systems are critical to monitor the melt pool behavior and control the deposition process in real-time. Optical sensor systems including IR cameras and OCT sensors allow prediction of melt pool width and height respectively, whereas pyrometers allow prediction of melt-pool temperature and behavior. Nevertheless, reliable control systems that are essential for real-time feedback control are still under development.

3.2. Research Needs

Based on the limitations identified from the literature review and information gathered from various studies and industrial surveys, the following research needs in terms of economical aspect and technological context have been formulated.

3.2.1. CAD/CAM Software Packages and its Limitations

A DVS study reported some of the key research areas identified in wire arc additive manufacturing process, from an industrial perspective, through expert interviews and questionnaire [128].

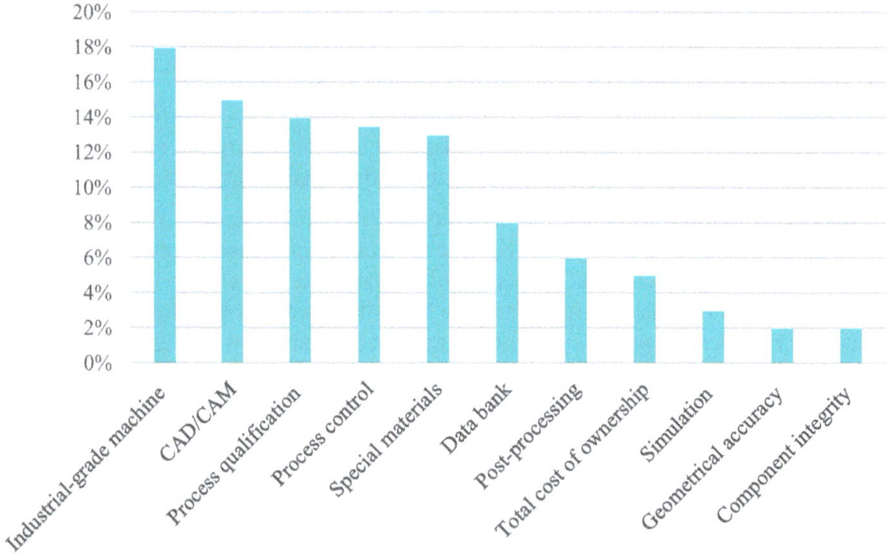

Figure 3.1 Key research areas in wire arc additive manufacturing process as identified through industrial surveys and interviews (based on [128])

Figure 3.1 shows the research areas identified through industrial survey. The study reported that 100 % of the surveyed industries did wish for software support for adoption and successful implementation of this technology, in which 55 % would prefer to use CAD/CAM software packages in combination with a CAE framework for thermo-mechanical simulation, while the other 45% were contended with CAD/CAM only software packages. Other reports also highlight the lack of automation in the existing software packages [129].The status-quo of the availability, functionality and their limitations of the CAD/CAM software packages or necessary algorithms for multi-axis depsotioin has been presented in section 2.5.

3.2.2. Higher Costs for Lot Size 1

Additive manufacturing has proven to be economical for metal prototyping applications across several industrial sectors. This is true for powder-bed based processes. However, for DED processes, the total cost per cu.cm of material deposition for lot size one is relatively higher in comparison to powder-bed processes. This is due to the high one-time engineering cost for toolpath planning, build strategy definition and the associated trial and error experiments (see Figure 3.2). But, the cost per cu.cm for a larger lot size is significantly lesser for both the AM technologies, while in case of DED technologies a drastic change can be witnessed due to its higher deposition rates [130].

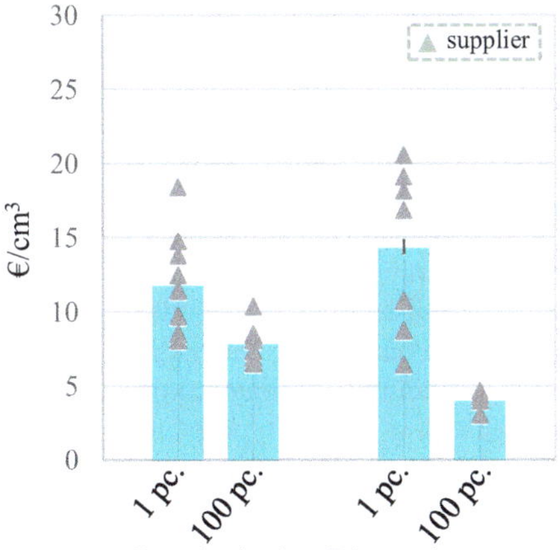

Figure 3.2 Overview of manufacturing costs for powder-bed and directed energy deposition processes with 1.4404 material from different suppliers (based on[130])

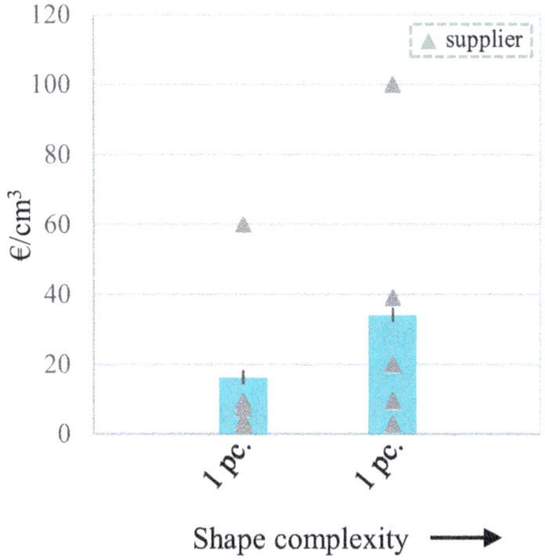

Figure 3.3 Overview of manufacturing costs for different directed energy deposition processes with Ti-6Al-4V material from different suppliers[1]

Figure 3.3 shows an overview of DED manufacturing costs once again normalized to €/cm³ for two different geometries with increasing geometrical complexity. It can be seen that the average cost increases with increasing shape complexity due to its associated toolpath planning and build strategy definition challenges. Also, the range of this normalized cost is higher in case of DED technologies due to the wide variety of available machine configurations and process types with machine costs ranging from 100 k€ to 2.0 Mn €.

[1] Refer Appendix A.1 for detailed cost anaylsis

3.2.3. Defects Caused by Higher Deposition Rates

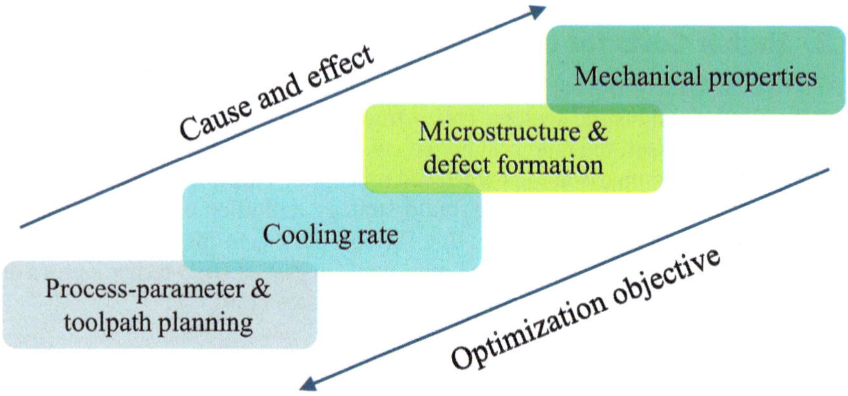

Figure 3.4 Cause and effect diagram of a directed energy deposition process (based on [37])

The resulting geometrical, metallurgical and mechanical characteristics of the DED component depends upon the cooling rate and the resulting microstructure of the individual layers (eventually of the individual weld tracks), see Figure 3.4 [37]. The cooling rate is essentially governed by several intrinsic and extrinsic process parameters which dictate the process mechanics or the melt pool dynamics. Interaction of primary process parameters (laser power, deposition velocity and material feed-rate) with secondary process parameters also influence the resulting metallurgical characteristics. Secondary process parameters include factors like component geometry, slicing technique, toolpath and build strategy, substrate geometry and temperature, clamping forces, type of nozzle, powder characteristics, heat treatment conditions and many others.

Higher deposition rates, in addition, gives rise to defects such as voids between boundary and hatch path, crater due to sub-optimal hatch distance, increased material accumulation at start/end points, intersections and corners [131–134]. High deposition rates being connected with higher energy input leads to formation of residual stresses, followed by deformation, delamination and hot cracks during the manufacturing process that needs to be addressed in the manufacturing planning phase [134–137]. The optimization objective in this case is primarily the correct choice of primary and secondary process parameters [138–140]. Figure 3.5 shows the commonly occurring defects in DED processes.

The common material handling system used in this layer-by-layer building process is either a 3-axis gantry system or a 6-axis industrial robot with an additional 2-axis turn-tilt table. The accuracy and control parameters of the material handling system has a direct implication on the geometrical characteristics of the built geometry in addition to the afore-mentioned causes [141]. Such parameters include acceleration and deceleration, interpolation parameter, motion type (linear or spline), etc., which have a direct dependency on the build quality.

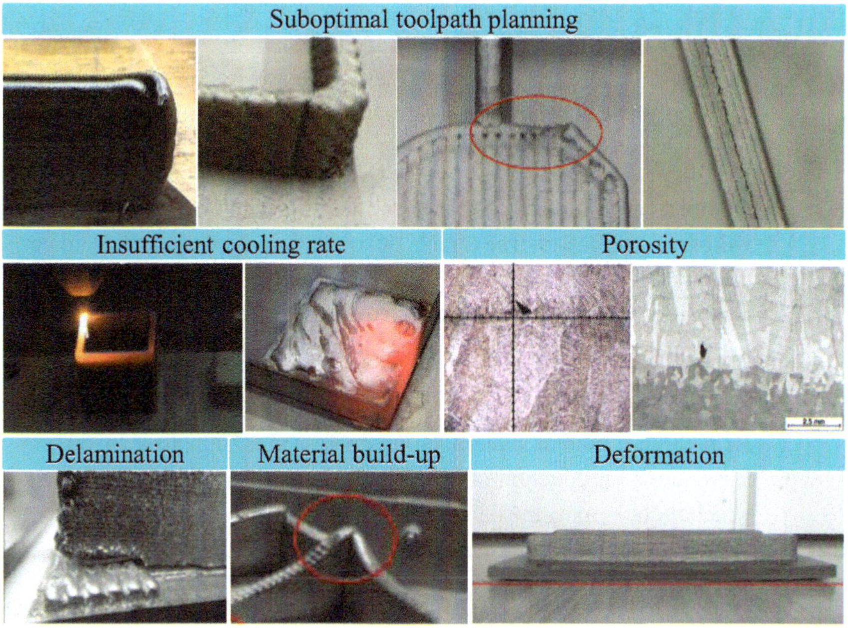

Figure 3.5 Common defects occurring in powder- and wire-based directed energy deposition processes

3.3. Research Goals

The research work carried out in this thesis focuses on the optimization objective represented schematically in Figure 3.4, namely process parameters and toolpath planning for high deposition rate additive manufacturing. The preliminary goal of optimization is to achieve a successful build job of aerospace components without toolpath-induced defects and with higher layer-wise geometrical accuracy that is critical for ensuring stable working distance throughout the build job. Based on the current state-of-the-art, research reports, industrial surveys and the inferred research gaps identified in section 3.1, research goals were determined that would contribute towards industrialization of the directed energy deposition technology. Research goals were classified under three important criteria, viz. cost, quality and time, and are tabulated in Table 3.1.

Table 3.1 Research goals classified under cost, quality and time

Cost	1. Reduction of **one-time toolpath planning** costs by 50% (min.) 2. Elimination of manufacturing **rejections** and the costs incurred with the trial-and-error experiments
Quality	3. Optimization of process parameters and toolpath strategy for deposition of **void-free** and **defect-free** (crack, delamination, binding defects, material accumulation and build failures) weld geometries 4. Deposition of layers with **higher geometrical accuracy** for complex three-dimensional components (± 1.0 mm) to ensure stable working distance along the build direction
Time	5. Reduction of time in manufacturing process planning phase by 50% 6. Elimination or reduction of time-intensive trial-and-error experiments, which reduces machine down-time

4. Research Methodology

Manufacturing planning phase of DED being highly component-geometry dependent, the research work begins with a clustering that classifies industrial use-cases based on their shape complexity and geometrical features and characteristics. Following a brief description of the clustering, the framework used for development and implementation of cluster-specific algorithms for pre-processing, slicing and toolpath generation is described. The developed algorithms are then elaborated in chapter 5. Chapter 6 presents a systematic experimental approach developed to identify primary and secondary process parameters for an industrial use-case. This chapter ends with the representation of a DED process chain in five phases that captures all the input and output system parameters with a focus on the manufacturing process planning phase.

4.1. Component Clustering

Several industrial use-cases were gathered from literature, press articles, research projects and also directly from industry partners. Based on the similarities in terms of geometrical features, slicing requirements, base surface and toolpath requirements, six different component clusters have been defined and tabulated in Table 4.1, along with CAD-model representation of demonstrator components for each cluster in Figure 4.1.

Table 4.1 Descriptive classification of typical DED use-cases

Bulk volume	Components where the volume of material to be deposited per layer is higher; Simple planar geometries; No special slicing or toolpath algorithms to be developed; Typically manufactured with boundary contour (with or without offsets) and internal volume fill patterns like raster. Equal distribution of heat needs to be ensured with appropriate wait time between layers; Distortion plays a major role that needs to be addressed with sufficient clamping.
Multi-directional	Components with overhangs or features that needs to be deposited in their normal orientation; Volume decomposition and feature segmentation is required before feature-wise slicing. Manufacturing restrictions based on the kinematic system used and the process parameters selected needs to be considered.
Gradually inclining	A special case of multi-directional.
Thin-wall	Typical aerospace structural components with strengthening elements like ribs and iso-grids. Usually planar slicing is sufficient.

V. J. Prakash, *Automated Process Planning for Additive Manufacturing with Directed Energy Deposition*, Light Engineering für die Praxis,
https://doi.org/10.1007/978-3-031-87666-0_4

	Rib intersections and corners are typical heat affected zones, where toolpath optimization is required which otherwise lead to geometrical inaccuracies and bonding defects.
Non-planar parametric	Non-planar additive components where base surface is one of the parametric surfaces like cylinder or sphere; Volume decomposition is often required which is achieved by segmentation based on Topology information; Non-planar slicing and toolpath generation algorithms are required.
Non-planar freeform	Non-planar non-parametric freeform components that are modeled as NURBS surfaces; Complex tool and die or aerospace components; In-addition to non-planar slicing algorithms, NURBS-based toolpath generation algorithms are required.

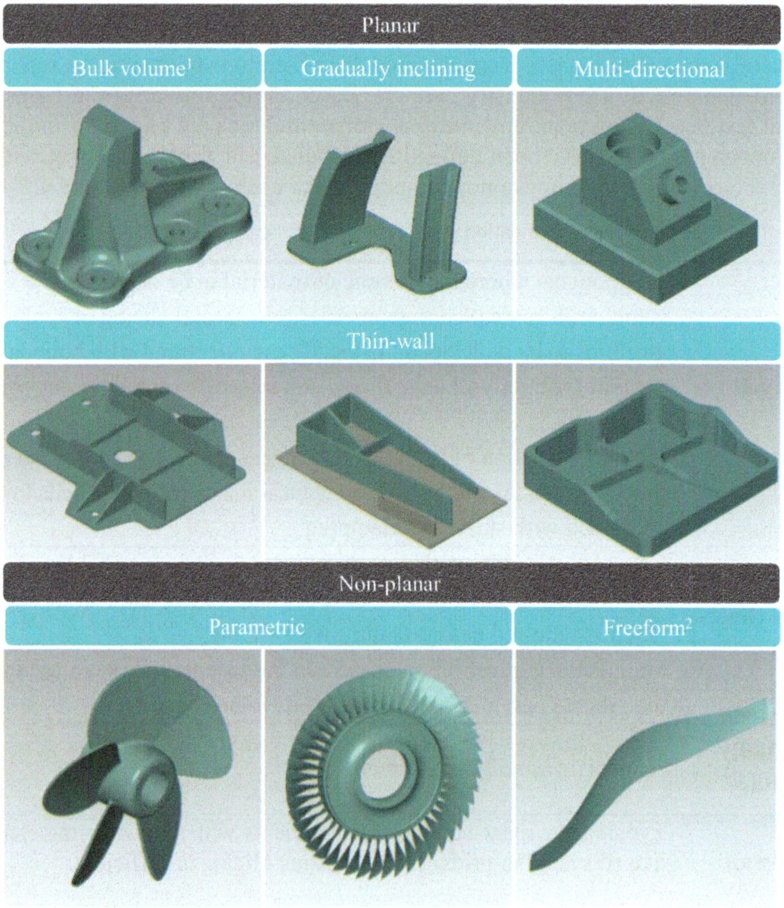

Figure 4.1 Clustering of DED demonstrator components with CAD model representation

4.2. Framework

This research work intends to improve the limitations of several algorithms and develop new algorithms with focus on design automation used for slicing and toolpath planning of DED components. For this purpose, a framework similar to that of a CAD/CAM software package is required. One possible approach is to use the application programming interface (API) of any such software package to extend the functionalities of its existing toolbox or develop new algorithms. Such an API approach would benefit from using the fundamental or existing functions and methods of the chosen platform. On the other hand, the algorithm or the implementation of it would be bounded by and restricted to the chosen software package. Therefore, a python-based framework with self-defined methods and open-source third-party libraries has been used (see Figure 4.2).

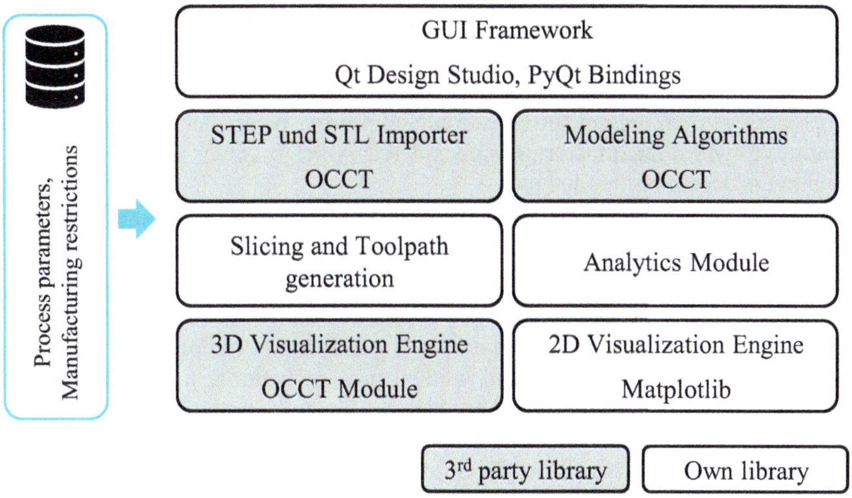

Figure 4.2 Framework used for development and implementation of toolpath generation algorithms

Two important third-party libraries were used for the implementation of the developed algorithms. One, the aforementioned open source Open Cascade Technology (OCCT) kernel[2] that enables computation of STEP-file format-based CAD models. Similar to STEP protocol, OCCT defines its own geometrical and topological entities that complies with the international standard, although differing in the naming scheme from that of STEP protocol (section 5.1and section 5.2). In some cases, like volume and feature analysis in section 5.3, STL-file format data has been used as an intermediate step owing to its simplicity and suitability with other libraries. The second third-party library is called NURBS-python (or geomdl), which has been used especially for toolpath planning of non-planar freeform components (section 5.4) [142].

[2] Open Cascade Technology geometry kernel was developed by Open Cascade, now a part of Capgemini Engineering

4.3. Process Chain

A general process flowchart of a directed energy deposition process has been charted that defined the process in five different phases, also capturing all possible individual steps and the influencing intrinsic system and process parameters (see Figure 4.3).

1. Concept phase: In this first phase of DED process, the design space for the geometry is specified along with the definition of constraints, loads, material, design variables and objective function. Powder-bed AM processes usually begins with a topology optimization process that spatially optimizes the material distribution within the design space satisfying the objective function, followed by additive manufacturing of the same. DED processes, however, do not provide the same degree of design freedom and are therefore restricted to size optimization.

2. Design phase: In the design phase, a CAD model with an over-mass for near-net shape deposition is developed. One of the most important factors to be considered in this phase are the manufacturing constraints and design guidelines, very specific to the chosen DED process, system architecture, material and process parameter set. Furthermore, the possibility of hybrid manufacturing is considered, which involves incorporating the substrate into the design of the final part.

3. Manufacturing planning phase: One of the essential phases in DED process chain, involving generation of deposition toolpath with component-specific algorithms and its optimization with thermal and kinematic simulation models is the planning phase. In the toolpath generation step, the entire CAD model is virtually 'sliced' according to defined layer thickness and toolpaths for every layer is generated. Furthermore, toolpaths can be optimized for minimum distortion and residual stresses, better heat distribution, constant deposition speed, productivity and other objective functions through validated simulation models. Process parameters like laser power, deposition velocity, material feed-rate, gas flowrates, etc. are then defined for the entire toolpath. CAD/CAM/CAE frameworks are used for the above-mentioned process steps. Optimized toolpaths are then exported into either G-codes or robot-readable programs using post processors.

4. Manufacturing phase: Whereas the first three phases take place in the digital layer of the process chain, the physical layer starts from this phase with the preparation of the substrate or base plate. The substrate is then fixed with the clamping system in the process chamber. The deposition system and its periphery (powder feeder, laser source and deposition nozzle) are calibrated in this phase for the deposition process. The deposition process is followed by heat treatment, measurement of geometrical deviation and final machining process.

5. Inspection phase: The manufacturing phase is followed by a final inspection phase, which includes final geometrical inspection for GD&T analysis and other quality inspections as required and defined by the standards.

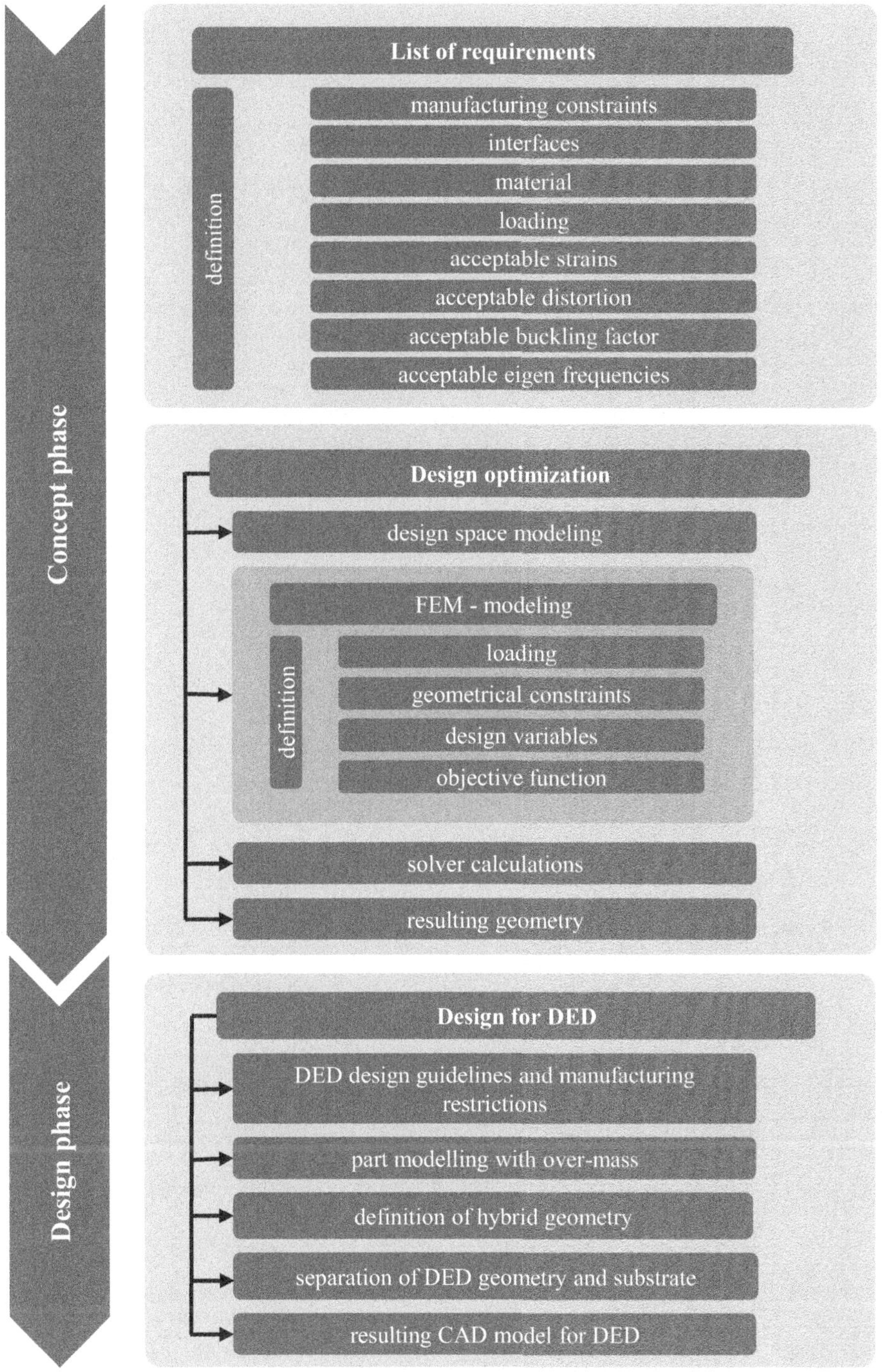
Concept phase
List of requirements
definition
manufacturing constraints
interfaces
material
loading
acceptable strains
acceptable distortion
acceptable buckling factor
acceptable eigen frequencies
Design optimization
design space modeling
FEM - modeling
definition
loading
geometrical constraints
design variables
objective function
solver calculations
resulting geometry
Design phase
Design for DED
DED design guidelines and manufacturing restrictions
part modelling with over-mass
definition of hybrid geometry
separation of DED geometry and substrate
resulting CAD model for DED

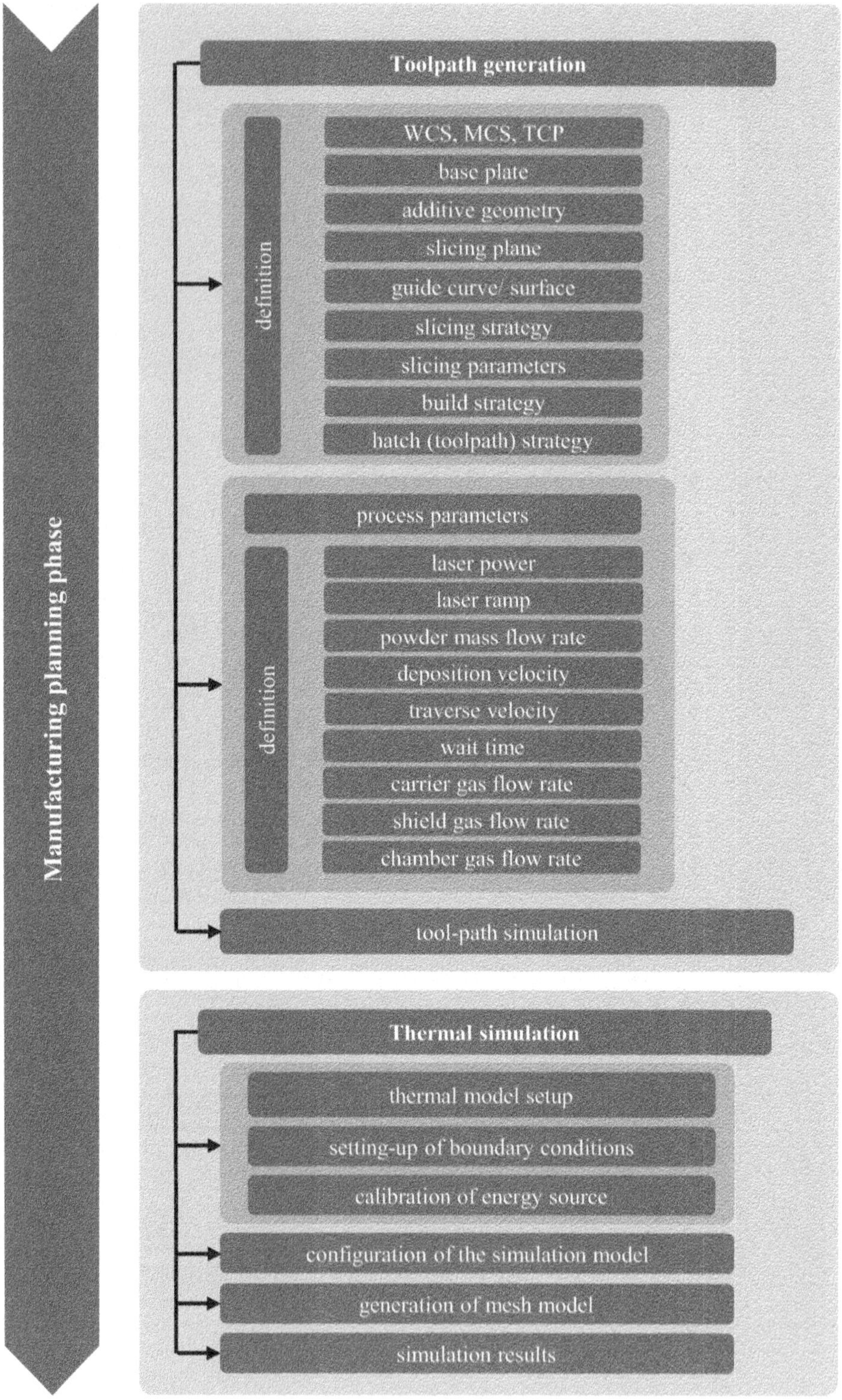
Manufacturing planning phase
Toolpath generation
definition
WCS, MCS, TCP
base plate
additive geometry
slicing plane
guide curve/ surface
slicing strategy
slicing parameters
build strategy
hatch (toolpath) strategy
process parameters
definition
laser power
laser ramp
powder mass flow rate
deposition velocity
traverse velocity
wait time
carrier gas flow rate
shield gas flow rate
chamber gas flow rate
tool-path simulation
Thermal simulation
thermal model setup
setting-up of boundary conditions
calibration of energy source
configuration of the simulation model
generation of mesh model
simulation results

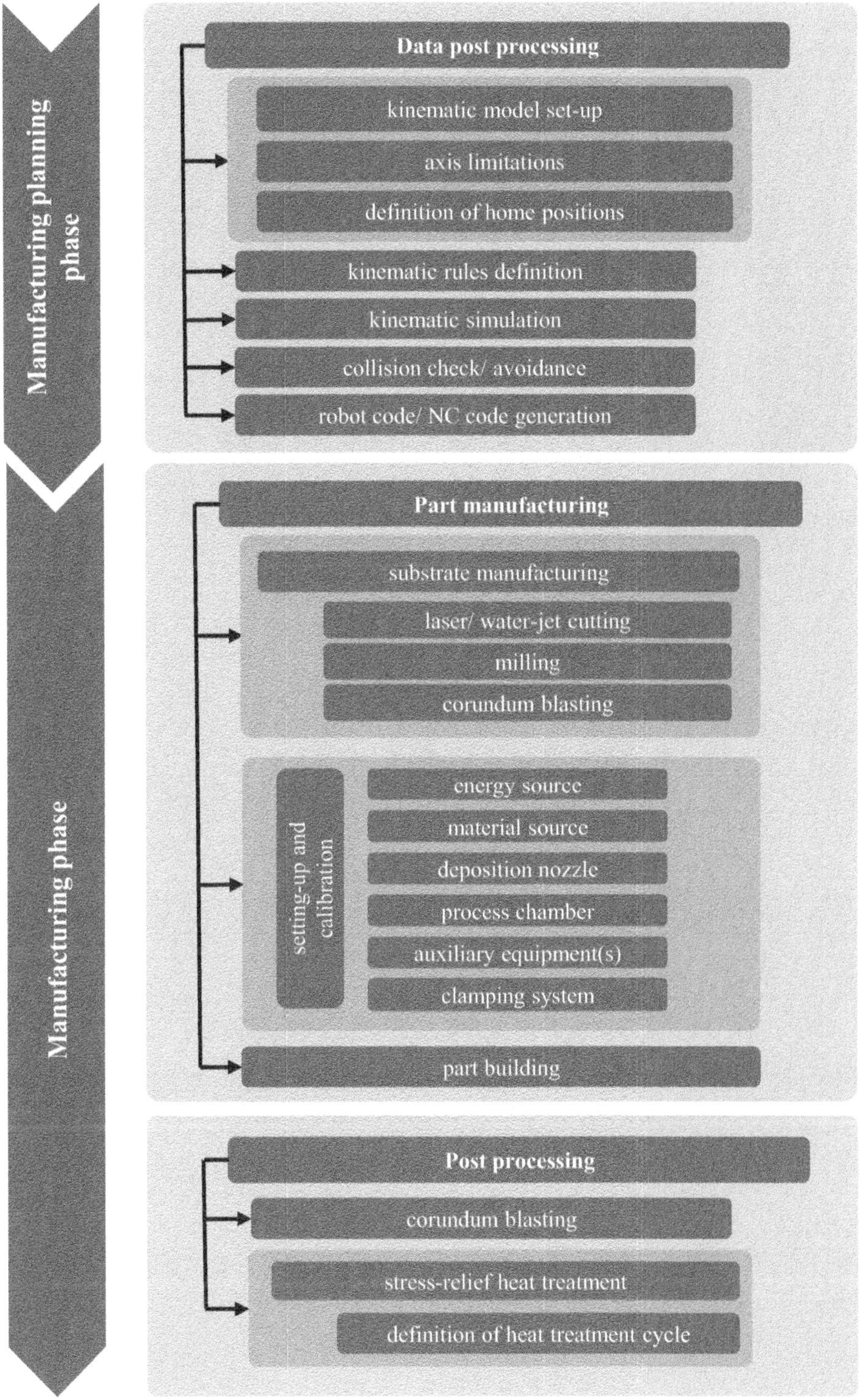
Manufacturing planning phase
Data post processing
kinematic model set-up
axis limitations
definition of home positions
kinematic rules definition
kinematic simulation
collision check/ avoidance
robot code/ NC code generation
Manufacturing phase
Part manufacturing
substrate manufacturing
laser/ water-jet cutting
milling
corundum blasting
setting-up and calibration
energy source
material source
deposition nozzle
process chamber
auxiliary equipment(s)
clamping system
part building
Post processing
corundum blasting
stress-relief heat treatment
definition of heat treatment cycle

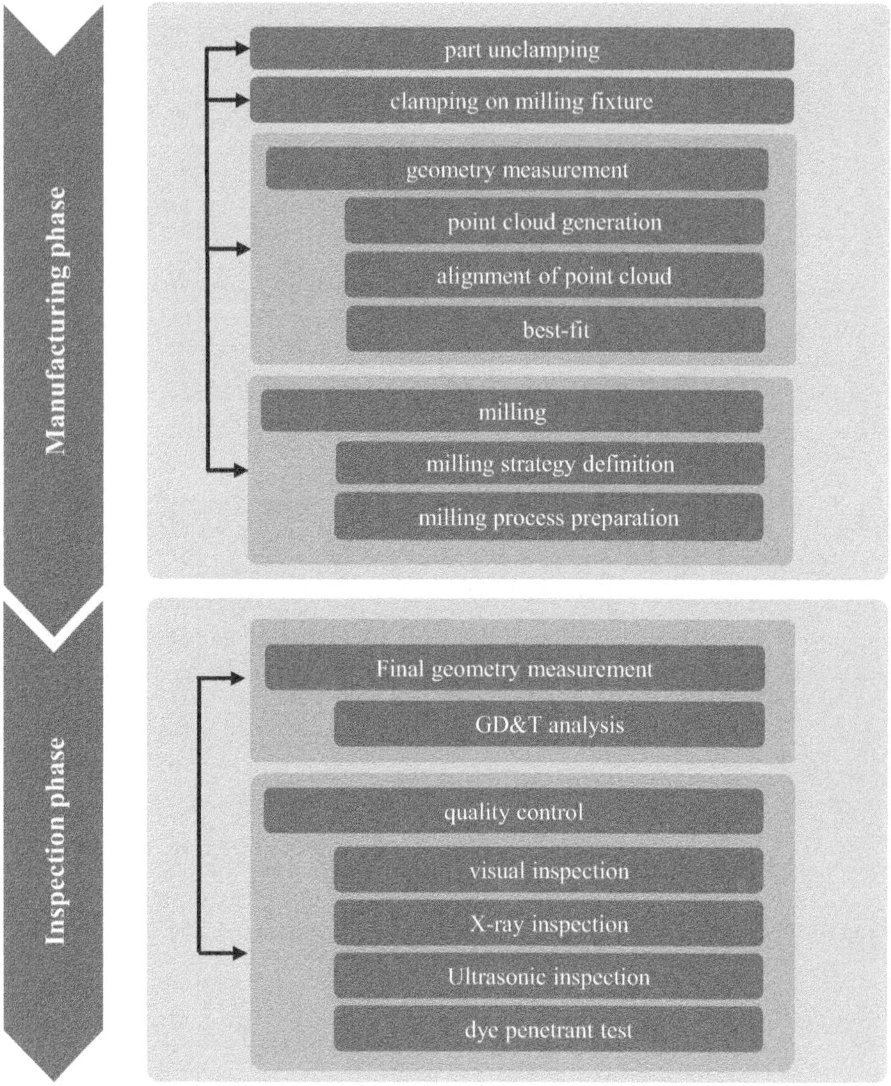

Figure 4.3 Process flowchart of a typical Directed Energy Deposition process

5. Slicing and Toolpath Generation Approaches

Algorithms, techniques and libraries necessary for 3-dimensional slicing and toolpath generation for robot-based DED processes are unique to the chosen geometry and its file format. This chapter presents such algorithms, pseudocodes and their implementation in Python. Results obtained for the chosen demonstrator components are also shown graphically.

5.1. STEP-File Format based Planar Slicing

For planar slicing of STEP-based CAD models, the aforementioned Open Cascade Technology (OCCT) kernel's geometric modeling algorithms have been used. Similar to STEP protocol, OCCT defines its own geometrical and topological entities that complies with the international standard though differs in the naming scheme. Unlike STL-file format which contains surface information as triangulated meshes, STEP exchange file format entails geometry information like points, vectors, directions, curves and surfaces as well as topological entities like vertices, edges, loops, faces, shells, volume and solids represented as mathematical constants. OCCT library encompasses the necessary classes for handling such datatypes along with several modelling algorithms, using which a slicer and a tool-path generation engine has been developed.

Data exchange module of OCCT, namely *STEP interface* module has been used to read STEP data that is in accordance with Application Protocol 214 and is translated into OCCT shape datatypes. Both geometrical and topological information can be extracted using this data exchange module. The *topology explorer* method can be used to break the STEP model into hierarchical topology entities. Table 5.1 shows the supported STEP topological entities with its equivalent OCCT shapes. To extract 3D and 2D geometrical entities, geometrical objects from the *Geom* package and *Geom2d* package have been used respectively. Table 5.2 shows some of the important class objects used in the development of this slicer engine along with their corresponding geometrical entity. An extensive list of all geometrical objects can be found in the documentation of OCCT [143].

Table 5.1 List of STEP-topological entities with its equivalent OCCT shape datatype

Topology	STEP Entity	OCCT Shape
Vertices	vertex_point	TopoDS_Vertex
Edges	oriented_edge, edge_curve	TopoDS_Edge
Loops	face_bound, face_outer_bound, edge_loop, poly_loop, vertex_loop	TopoDS_Wire
Faces	face_surface, advanced_face	TopoDS_Face
Shells	connected_face_set, oriented_closed_shell, closed_shell, open_shell	TopoDS_Shell

V. J. Prakash, *Automated Process Planning for Additive Manufacturing with Directed Energy Deposition*, Light Engineering für die Praxis,
https://doi.org/10.1007/978-3-031-87666-0_5

Table 5.2 List of STEP-geometrical entities with its equivalent OCCT objects

Geometry	STEP entity	OCCT Object
Points	cartesian_point	Geom2d_CartesianPoint
Directions	direction	Geom2d_Direction
Vectors	vector	Geom_VectorWithMagnitude
Curves	circle	Geom2d_Circle, Geom2d_BSpline-Curve
	line	Geom_Line, Geom2d_Line
	pcurve	Geom2d_Curve
	offset_curve_3d	Geom_OffsetCurve
	b_spline_curve, b_spline_curve_with_knots, bézier_curve, rational_b_spline_curve	Geom_BSplineCurve, Geom2d_BSplineCurve
	surface_curve	TopoDS_Edge
	boundary_curve	TopoDS_Wire
Surfaces	b_spline_surface, b_spline_surface_with_knots, bézier_surface, rational_b_spline_surface	Geom_BsplineSurface
	cylindrical_surface	Geom_CylindricalSurface

For planar slicing of a CAD geometry, a slicing surface and a slicing vector is required along with the geometry (shape entity) that needs to be sliced. The foundation face could be any user-defined surface that has been extracted using the *topology explorer* method. In case of planar slicing based on principal axes, the required axis information is taken as an input, with which a plane (geometry) and a face composing this plane (topology) is generated. Passing *RepAlgoAPI_Common* class with two arguments, namely, the topology solid entity to be sliced and the foundation face, will result in the intersection of two entities, which eventually is the sliced surface (a topological face entity). The slicing surface is then translated along the normal vector direction by a distance equal to layer thickness. The slicing process is repeated once again to generate the intersection face. The process continues until the function yields a null value, in which case the entire solid will be sliced. The slicing function is followed by generation of toolpaths (boundary offset paths and hatch paths) for each and every sliced surface. Boundary offsets can also be generated for every sliced surface using the function *self.offset_from_face* with two arguments, namely, the sliced face and a boundary offset distance. To generate raster-type toolpaths within a sliced surface, the very same *RepAlgoAPI_Common* class with two arguments, namely, the sliced surface and a hatching surface has been used, which returns an intersection curve (edge entity). The hatching surface is translated along the hatch vector by a distance defined by the hatch distance. The sliced surface now entails contour or boundary information represented as loops and hatch information represented as edges.

Following the hatching function, toolpath points necessary for robotic program creation is determined by evaluating points on the extracted loops and edges in the sliced surface. The extraction of points is done with the function *GCPnts_UniformAbscissa,* with two arguments namely, the edge or loop and an integer which define the distance between two evaluation points measured on the principal axis. Normal vector information at each evaluation point can be extracted using the function *curve.D0.* Pseudocode for the above-mentioned slicing, hatching and toolpath point extraction is written below.

5.1.1. Pseudocode

Function:	Planar slicing and hatching of a STEP shape entity
Input:	STEP-file, slice vector, slice thickness, boundary offset distance, hatch pattern, hatch vector, hatch distance
Result:	Returns nested list of co-ordinates with translation and rotational elements segregated between boundary path(s) and hatch path(s) for all layers

```
Read step file using OCCT ReadFile() function
Extract topology entities from STEP-file by iterating through OCCT TopoDS_Iterator()
    function
Generate slicing_plane (geometry) using gp_Pln(self.origin, gp_Dir(r⃗))
Convert slicing_plane into slicing_face (topology) using BRepBuilderAPI_MakeFace()
#slicing function
while True:
    extract sliced_surface using the method
    BRepAlgoAPI_Common(solid, slicing_face)
if (not sliced_surface.HasGenerated()) && (not sliced_surface.HasModified()) &&
    len(self.slices) ! = 0:
        break
    offset slicing_plane using slicing_plane.Translate(slice_thickness)
#boundary offset function
for slice_face in slices:
    generate offset boundary using
    self.offset_from_face(sliced_surface, boundary_offset_distance)
#raster hatching function
while True:
    create hatching_plane using gp_Pln(self.origin, gp_Dir(r⃗))
    hatch_section = BRepAlgoAPI_Section(sliced_surface, hatching_plane)
  if hatch_section.HasGenerated():
        weldpath_compound = hatch_section.Shape()
        exp = TopologyExplorer(weldpath_compound)
        weldpath_edges = list(exp.edges())
        for edge in weldpath_edges:
```

```
            intersection.append(edge)
        elseif hatch_section.HasModified():
            intersect_checker.append(True)
    else:
            if len(intersect_checker) == 0 && len(intersection) == 0:
                counter = counter + 1
    if len(intersect_checker) > 0 || len(intersection) > 0:
            if (not hatch.HasGenerated()) && (not hatch.HasModified()):
                break
    hatching_plane.Translate(manipulation)
#evaluation of boundary points and hatch points
for i = 1 to length(slices)
extract slice_wires from slices using TopologyExplorer(self.slices[i])
    for wire in slice_wires:
      extract wire_edges from Exp_slice.edges_from_wire(wire)
      for edge in wire_edges:
          generate curve using BRepAdaptor_Curve(edge)
          extract points on curve using GCPnts_UniformAbscissa(curve, 1.0)
      using curve.D0(ua.Parameter(n), p) extract translation and rotational vectors of
    the point
repeat for evaluation of hatch points
```

5.1.2. Implementation Results

The developed algorithm has been implemented in a python environment. Figure 5.1 shows the implementation result of the planar slicing algorithm on an aerospace demonstrator component. The demonstrator is a doorstopper typically used in the cabin doors of an aircraft. Figure 5.1.a shows the STEP-file rendering of the demonstrator component. Figure 5.1.b and Figure 5.1.c shows graphically the slices generated along the slice vector +z. Figure 5.1.d shows the hatch paths generated along the hatch vector +x along with the toolpath points extracted on the generated boundary paths and hatch paths.

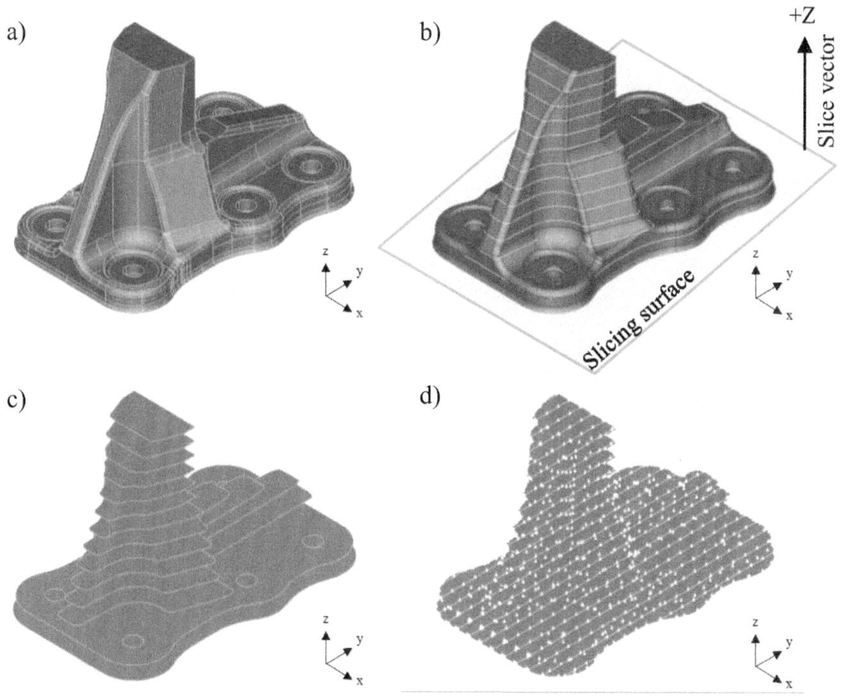

Figure 5.1 Planar slicing of STEP-file based CAD geometry: a) Solid STEP entity, b) Planar slicing along the slice vector, c) Sliced surfaces, d) Extracted toolpath points on the boundary and hatch toolpaths

Figure 5.2 shows the sliced surface with a hatching plane and a hatch vector along which the hatch paths are extracted. Sliced surface is a face topological entity which is composed of several boundary loops. Boundary offsets can be generated for every loop, however, is not shown graphically in this demonstrator.

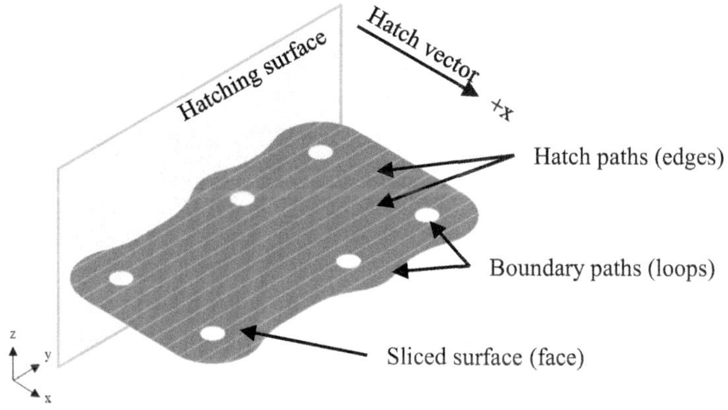

Figure 5.2 Graphical representation of the sliced surface with boundary paths and hatch paths generated along the hatch vector

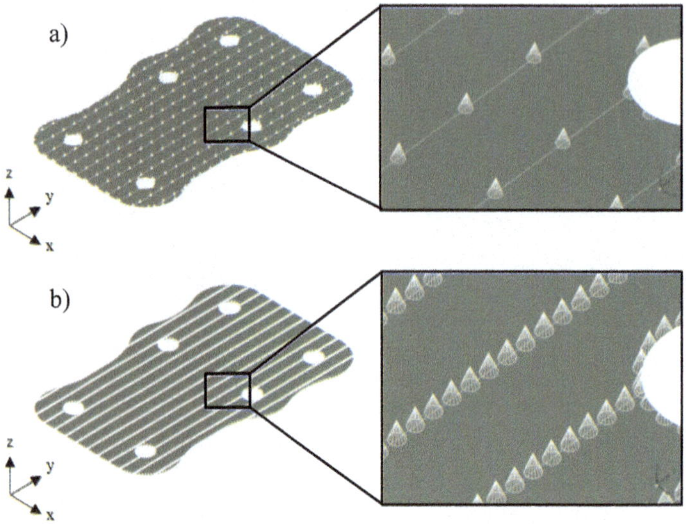

Figure 5.3 a) Toolpath points and normal vector extracted at a distance of 5 mm from each other b) Toolpath points and normal vector extracted at a distance of 1 mm from each other

In the STEP-file format-based toolpath generation, the sliced surface is represented in terms of curves enclosing the surface, which in-turn is represented mathematically. This provides the possibility to use computational geometry algorithms to extract toolpath points along the curve and also enable the extraction of normal and tangential vectors on a particular point using the first derivative function. The flexibility to choose the distance between two evaluation points can be leveraged to find a balance between shape complexity and total number of data points which decides the size of the robot program. Figure 5.3 shows two scenarios where the evaluation points are extracted from boundary and hatch paths at a distance of 5 mm and 1 mm for the same sliced surface. The change can be made possible by simply adjusting a variable that is passed as an argument to a function. In case of STL-file format, different files with different tessellation resolutions would be required for this purpose. Alternatively, an additional interpolation function would be necessary to extract evaluation points between the two points on the intersected tessellation surface, which will still not represent the original geometry accurately. This shows the advantage of working with STEP-file format rather than the industry de-facto STL-file format.

Figure 5.4 shows the representation of the same demonstrator geometry in two different file formats, namely, STEP-file format and STL-file format. The insert shows a magnified portion of the demonstrator, which clearly differentiates between the representation of surface information as a set of boundary curves definable with higher order mathematical equations and tessellated mesh geometry without topological information.

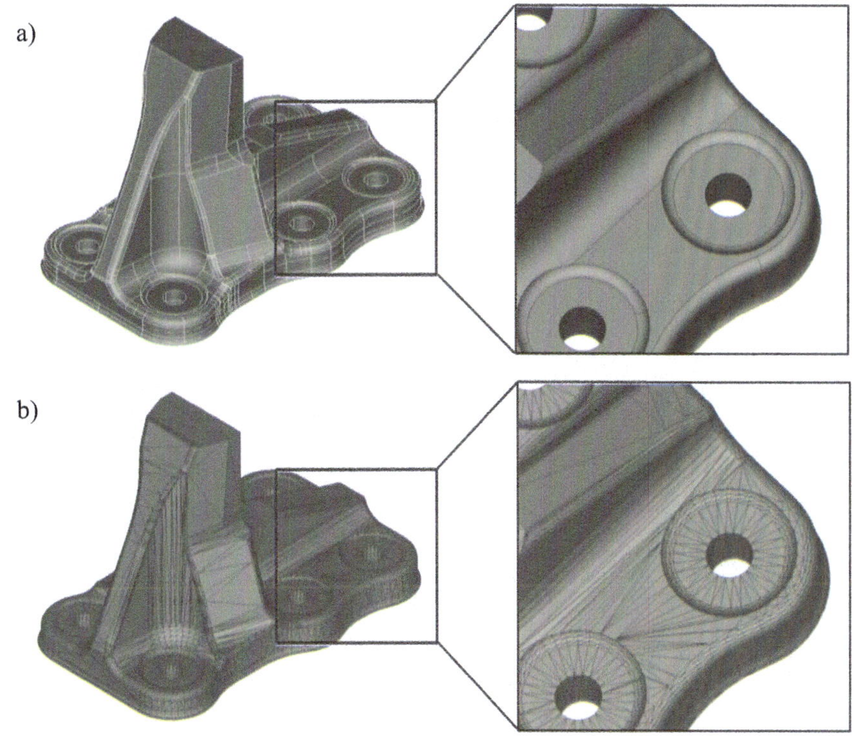

Figure 5.4 a) Representation of the demonstrator component in STEP-file format b) Representation of the demonstrator component in STL-file format

5.2. STEP-file Format based Non-planar Slicing

The approach to non-planar slicing is similar to that of the planar slicing described in section 5.1, except for the type of the slicing surface and the hatch surface used, which are in this case non-planar. This section explicitly deals with slicing of parametric non-planar solids and surfaces, namely cylindrical, conical and spherical surfaces. Slicing and hatching of volumes with freeform surfaces, namely B-spline and Bézier surfaces is discussed in detail in section 5.4 of this chapter. Non-planar slicing is frequently required for fabrication of cylindrically symmetric components like ship propellers and turbine blisks (portmanteau of bladed disks), which are some of the best use-cases for DED technology. In such components, the hub is usually manufactured by conventional routes like forming and acts as a substrate for deposition of blade geometries through one of the DED processes.

From data pre-processing perspective, the demonstrator components shown in Figure 5.5 needs to be initially decomposed into two different volume groups, namely, substrate and AM geometry, prior to the slicing and toolpath generation step. This volume segmentation step usually requires user intervention to manually identify and classify substrate from the additive geometry. This manual segmentation step is time consuming and thereby increases the engineering cost involved. The added cost and time can be negligible in case of serial production, however, for prototyping and lot size 1 production, the engineering costs incurred could be higher than the fabrication costs. Therefore, several algorithms have been developed to automate the volume segmentation step (tabulated in Table 2.1). Most of the algorithms focus on the development of volume segmentation approaches for STL-file format-based geometries. However, this section of the chapter deals with STEP-file format geometries. Volume segmentation carried out using OCCT library methods is presented in the following section, followed by non-planar slicing and toolpath generation.

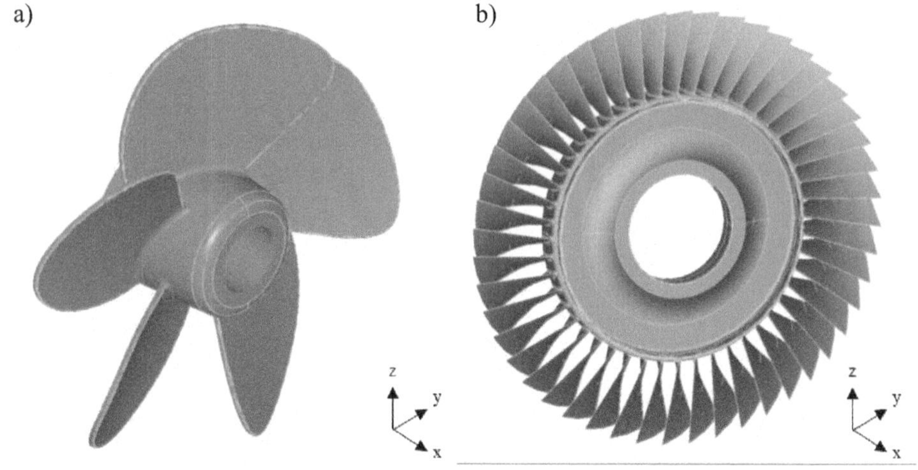

Figure 5.5 Demonstrator components for non-planar slicing based on STEP-file format: a) Propeller geometry b) Blisk geometry (not-to-scale)

5.2.1. Topology based Volume Segmentation

The fact that the STEP-file format carries not only geometrical information, but also topological information can be leveraged directly to achieve volume segmentation. Using OCCT's *TopologyExplorer* function, a compound solid can be decomposed into several solids, every solid into several surfaces, and every surface into several wires and wire into edges. Volume segmentation is essentially a topology explorer. In the demonstrators shown in Figure 5.5, the hub must be separated from the propeller blades in both the cases. The hub acts as a substrate (and thereby slicing surface or foundation surface), whereas the blades of the propeller and blisk geometry are shape entities used for slicing and toolpath generation. Figure 5.6 shows the demonstrators decomposed into several STEP topological entities like plane, torus, cylinder, cone and B-spline surface using the *TopologyExplorer* function. The hub of the ship propeller is composed of cones and tori, whereas the blade geometry is composed of cylinders and B-spline surfaces. The extracted topological entities are assigned to either the list base_surface or to the list slicing_surfaces. In case of the ship propeller geometry, cones and tori are assigned to base_surface, whereas the rest of the entities to slicing_surfaces. This eliminates the need of manual selection of individual features or entities by the user. Similarly, in case of the blisk geometry, all B-spline entities constitute to the blade geometry and therefore can be assigned to slicing_surfaces, whereas the rest of the entities to base_surface. This is one step towards automation of volume segmentation, however, still relies on user intervention to group certain topological entities together from the programing level rather than the graphical interface level.

Figure 5.7 shows the results of a topology explorer function on the doorstopper demonstrator component chosen in the previous section. It is interesting to note the number of different topology entities that constitute the demonstrator. The figure also shows the count of several topological entities extracted from the demonstrator.

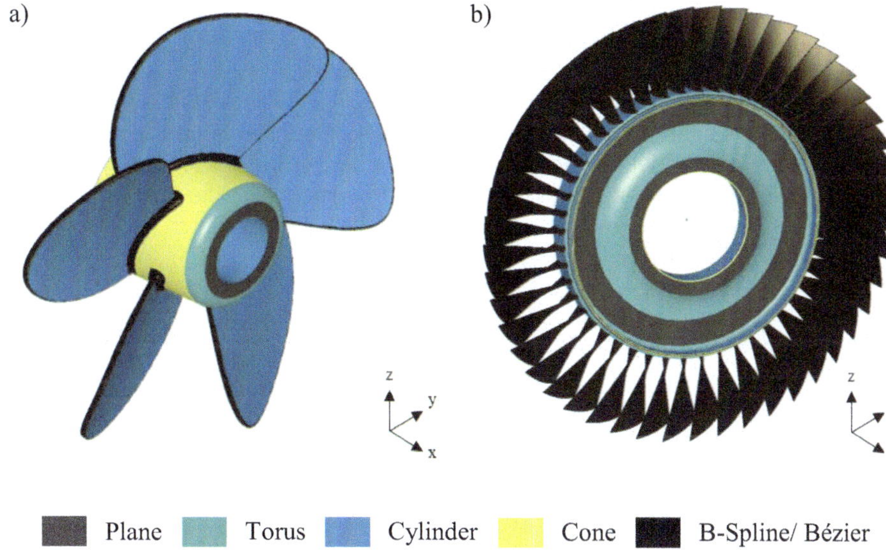

Figure 5.6 Demonstrator components decomposed into several topological STEP entities using OCCT topology explorer function

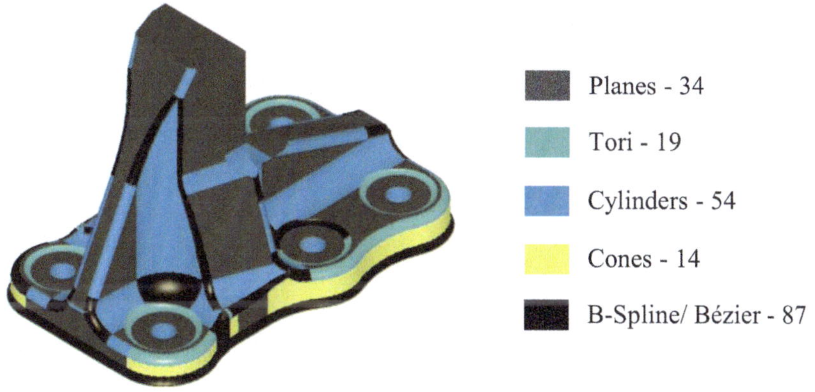

Figure 5.7 Topology explorer applied for doorstopper demonstrator component showing also the count of different topology entities

5.2.2. Pseudocode

Following the volume segmentation between base_surface and slicing_surfaces, non-planar slicing and toolpath generation will be carried using the same *BRepAlgoAPI_Common* function with two arguments, namely, the outermost face of the base_surface and the slicing_surfaces. The function used remains the same as the function used for planar slicing, it differs only on the shape entity of the base_surface being used. In the case of demonstrator components shown, it will be either cylinder or conus on the propeller hub. After successful slicing of the first slice, the base_surface is translated radially by a distance equal to the slicing thickness. Toolpath generation or hatching is carried out on the sliced faces by either offsetting the boundary paths from outside to inside or by generating additional hatch paths by an intersection of a hatch surface with the sliced surface. The pseudocode for volume segmentation, non-planar slicing and toolpath generation using OCCT classes and functions is described below.

Function: Volume segmentation, non-planar slicing and hatching of a STEP shape entity

Input: STEP-file, slice surface, slice vector, slice thickness, boundary offset distance, hatch pattern, hatch vector, hatch distance

Result: Returns nested list of co-ordinates with translation and rotational elements segregated between boundary path(s) and hatch path(s) for all layers

```
Read step file using OCCT ReadFile() function
#volume segmentation function
Extract topology entities from STEP-file using OCCT TopologyExplorer() function
Iterate through faces and save different lists corresponding to surface type
Define base_surface from the extracted surface list constituting to the hub
Define slicing_surfaces from the extracted surface list that constitute to blade geometry
#slicing function
while True:
    extract sliced_surface using the method
    BRepAlgoAPI_Common(slicing_surfaces, base_surface)
if (not sliced_surface.HasGenerated()) && (not sliced_surface.HasModified()) &&
    len(self.slices) ! = 0:
        break
    offset base_surface radially using base_surface.SetRadius (radius)
    radius += slice_thickness
#boundary offset function
for slice_face in slices:
    generate offset boundary using
    self.offset_from_face(sliced_surface, boundary_offset_distance)
#mid-line hatching function
```

```
while True:
    define hatching_plane from the cylinder shape entity of the propeller blade
    geometry
    hatch_section = BRepAlgoAPI_Section(sliced_surface, hatching_plane)
  if hatch_section.HasGenerated():
        weldpath_compound = hatch_section.Shape()
        exp = TopologyExplorer(weldpath_compound)
        weldpath_edges = list(exp.edges())
        for edge in weldpath_edges:
          intersection.append(edge)
    elseif hatch_section.HasModified():
        intersect_checker.append(True)
  else:
        if len(intersect_checker) == 0 && len(intersection) == 0:
          counter = counter + 1
  if len(intersect_checker) > 0 || len(intersection) > 0:
        if (not hatch.HasGenerated()) && (not hatch.HasModified()):
          break
  hatching_plane.Translate(manipulation)
#evaluation of boundary points and hatch points
for i = 1 to length(slices)
extract slice_wires from slices using TopologyExplorer(self.slices[i])
    for wire in slice_wires:
      extract wire_edges from Exp_slice.edges_from_wire(wire)
      for edge in wire_edges:
          generate curve using BRepAdaptor_Curve(edge)
          extract points on curve using GCPnts_UniformAbscissa(curve, 1.0)
      using curve.D0(ua.Parameter(n), p) extract translation and rotational vectors of
    the point
repeat for evaluation of hatch points
```

The pseudocode has been implemented once again in a python environment and the results are graphically presented in the next section.

5.2.3. Implementation Results

For the sake of better visualization, the demonstrator geometry is simplified to have only one blade geometry. The substrate and additive geometry are segmented and differentiated by color (see Figure 5.8.b). Figure 5.8.c and Figure 5.8.d show the non-planar slicing surfaces and non-planar sliced surfaces respectively.

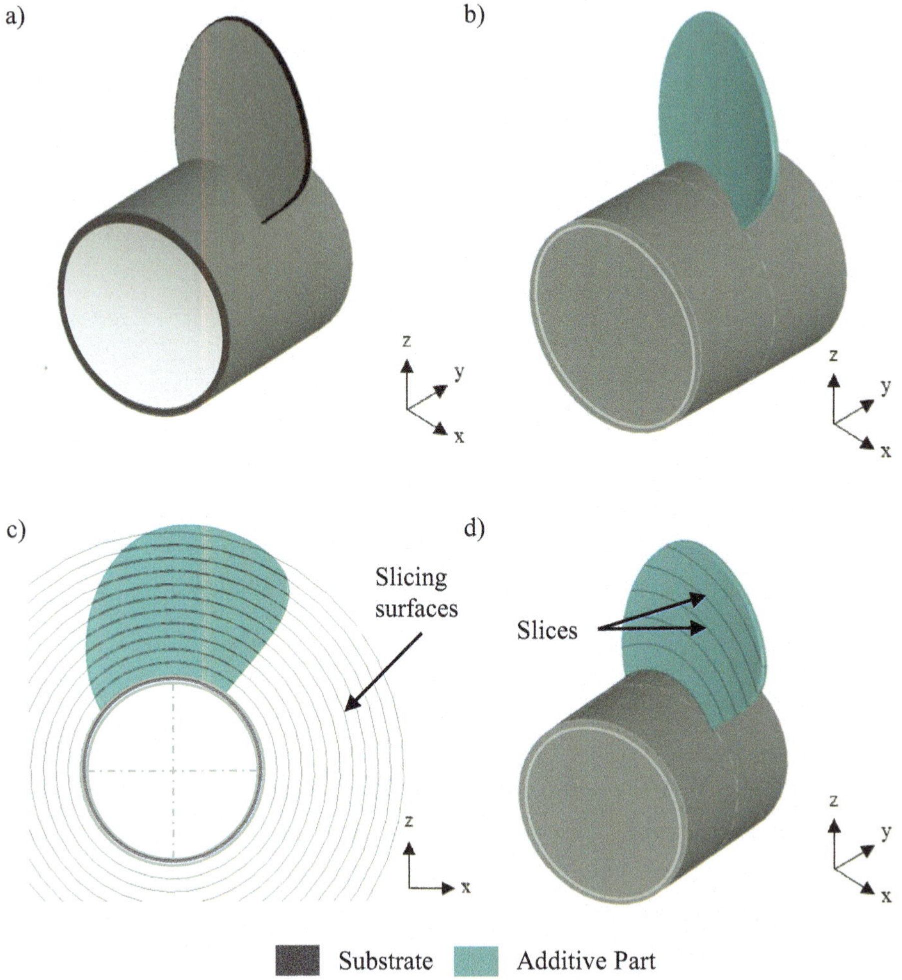

Figure 5.8 a) Simplified demonstrator propeller geometry for non-planar slicing b) Volume segmentation of base surface and slicing surfaces c) Visualization of slicing surfaces translated radially outwards d) Visualization of non-planar slices on the propeller blade geometry

Figure 5.9 shows an enlarged view of the sliced non-planar surfaces extracted from the demonstrator component. Hatch toolpath in addition to boundary toolpath has been generated by intersecting the sliced surface with a hatch surface, defined by translating the cylindrical surface of the blade geometry radially inwards. Additionally, points have been evaluated on both the toolpaths along with the extraction of normal vectors.

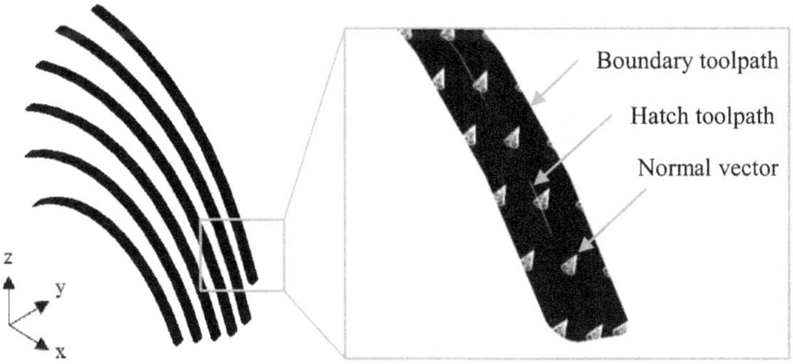

Figure 5.9 Graphical representation of non-planar slices of the propeller blade geometry with an enlarged view showing normal vectors of the evaluated points on the boundary and hatch toolpaths

5.3. Multi-directional Slicing

Data pre-processing steps required for this component cluster also requires recognition of features and segmentation in the build direction before the implementation of slicing and hatching algorithm. The difference in the volume segmentation of this component cluster in comparison to section 5.2.1 is the fact that the segmentation takes place within the additive part geometry itself, rather than between the substrate and additive part. OCCT's topology explorer function could be used once again for STEP-file based CAD models to differentiate between several topological entities of the additive geometry. However, in this section, the most commonly used STL-file format has been chosen and algorithms for automation of volume segmentation have been developed. For this purpose, centroidal axis-based feature extraction has been used owing to its simplicity.

5.3.1. Centroidal-axis based Volume Segmentation

Extraction of geometrical topology information and feature hierarchy based on centroidal axis information is carried out based on the evaluation of the following three geometrical parameters.

Centroidal Axis Extraction

Centroidal-axis is one of the many geometrical parameters to deduce the topological information and geometrical continuity of a STL-file format-based CAD model in a reliable manner [59]. In the principal build direction (Z-axis), centroids (C) obtained at a regular interval results in a list of centroids forming Centroidal-Axis (CA). The connection between adjacent centroids is called Link (L). Centroidal-Axis therefore comprises of a set of centroids and associated links. The mathematical formulation of the Centroidal-Axis can be expressed using set theory as follows.

$$CA = \{C_1, L_{12} \dots . L_{1k} | \dots | C_i, L_{i1} \dots . L_{ik} | \dots | C_n, L_{n1} \dots . L_{nk}\} \quad (5.1)$$
$$i = 1 \dots . n$$
$$L_{ij} = < C_i, C_j > \quad i \neq j$$

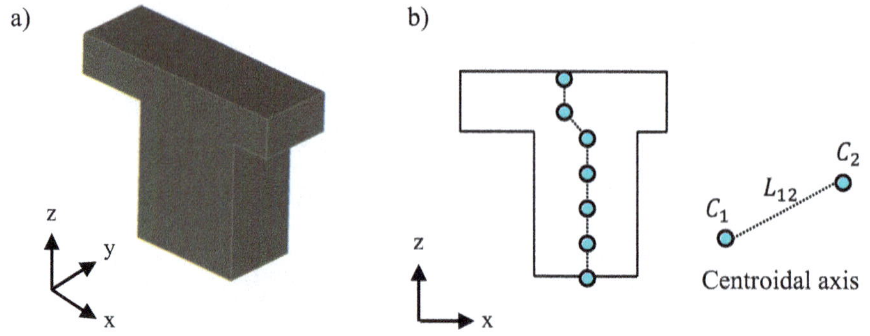

Figure 5.10 a) STL-file based demonstrator component with overhangs; b) Centroidal-axis extracted for the demonstrator geometry

Figure 5.10 shows the centroidal-axis computed along +z direction for a simple geometry with two overhanging features protruding along the +x and -x direction. A shift in the centroidal-axis as seen in Figure 5.10.b can help determine the presence of an overhang or geometry change. A shift along the negative x- direction indicates a geometry change in this direction and this information is used for volume segmentation and determination of a new slicing plane. However, the geometry contains a second overhang in the positive x- direction which is left undetected due to its relative size with the overhang in the negative x- direction. Therefore, an additional parameter is required for a holistic determination of features or overhangs, which is carried out through successive cross-section validations.

Validity of cross-section

Two successive cross-sections, lower plane (A_l) and upper plane (A_u) in the build direction, are obtained from the intersection planes. The upper cross-section is projected onto the lower to obtain one of the following three cases after computing $A_u - (A_u \cap A_l)$:

Case 1: A_u

Case 2: Φ

Case 3: $\Delta A_1, \Delta A_2, \Delta A_3, \ldots, \Delta A_n; \sum_{k=1}^{n} \Delta A_k < A_u$

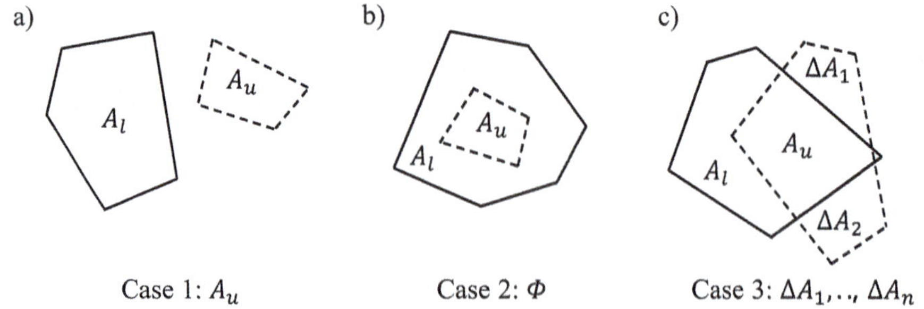

Figure 5.11 Three possible cases obtained during the validity of cross-section between two consecutive intersection planes

Case 1 implies that the upper cross-section does not intersect lower cross-section, thereby concluding that there is no connectivity in geometry. Whereas in case 2, the upper cross-section is a subset of lower cross-section and hence there is no indication of an overhang or a sudden change in geometry. Both the above-mentioned cases eliminate the need for centroidal axis extraction of the branch. In case 3, there would exist protruding areas that may be overhangs which suggests a change in direction of slicing plane. These could be segments that require slicing in a new plane or it could also be geometrical features like fillets and chamfers or a gradually inclined geometry. In the latter case, these features must not be segmented and could be built in the principal build direction unless exceeding the maximum overhang permitted by the process. In order to identify the correct protruding area as a segment, a user-defined overhang area parameter (A_o) is chosen which is defined as:

$$A_o = {}^{\Delta A_i}/_{A_u} * 100 \tag{5.2}$$

Selection of a protruding area to check the change in direction condition is done if it crosses the predefined threshold value of A_o. In case of the representative scenario shown in Figure 5.11.c, two protruding areas namely ΔA_1 and ΔA_2 occur, thereby identifying two overhang structures.

Determination of direction change

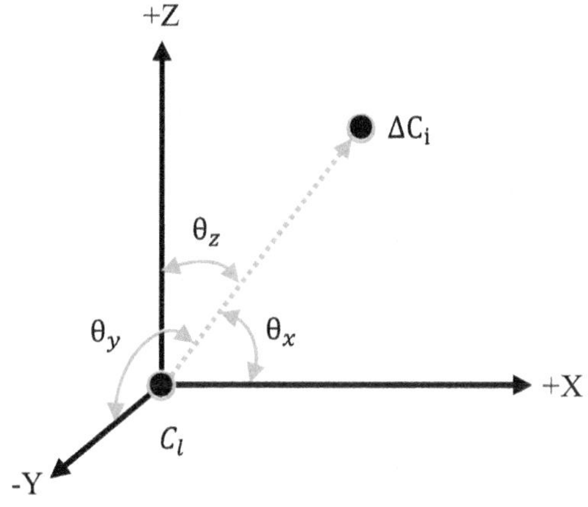

Figure 5.12 Schematic representation of the link element between the centroid of the lower plane and the centroid of the overhanging area and its corresponding angle with three principal axes

Centroid for every feature $\Delta A_1, \Delta A_2, \ldots, \Delta A_n$ is determined as $\Delta C_1, \Delta C_2, \ldots, \Delta C_n$ respectively. A feature could be an overhang, branch or an inclined surface of a gradually inclined component. Vectors joining each of these centroids with the centroid of the lower cross-section CA_l are constructed. Angles between these vectors with all principal axis, namely X, Y and Z are measured, as shown in Figure 5.12. For each ΔA_n, the smallest angle is selected and the principal direction with which the vector forms the smallest angle is selected. If this specific angle, overhang angle (α) exceeds certain threshold compared to the angle formed by the same vector with the current principal direction, then such principal direction is the new plane normal for an overhang represented by that area. The threshold value is defined by the process restriction which is the maximum wall inclination possible to build without tilting the deposition head, for a given DED system setup.

Algorithm summary

Based on the above-mentioned three geometrical parameters, centroidal axis for a geometry along with its branches are determined. Every centroidal-axis is then treated as a segment and can be sliced with a different slicing vector indicating the direction of the segment. For any given geometry, the initial direction of normal ($\vec{r}$) for intersection is taken along the commonly chosen +z direction and the first intersection plane at (0, 0, z_{min} + 0.001). The first intersection plane is computed along with the computation of the centroid as C_n and area as A_n. The next cross-section plane for evaluation is defined as:

$$S_{n+1} = S_n + (step.\vec{r}) \tag{5.3}$$

where,

$\vec{r} \in \{[1,0,0], [-1,0,0], [0,1,0], [0,-1,0], [0,0,1], [0,0,-1]\}$

$step \leq$ slicing thickness

Plane intersection of the current branch at S_{n+1} is computed along with its centroid as C_{n+1} and area A_{n+1}. Two successive cross-sections are evaluated for condition $A_{n+1} - (A_{n+1} \cap A_n)$ and one of the three cases mentioned above will be executed. Case 3 indicates new segment, that is a child branch of the current branch. The change in direction of intersection plane normal is determined for the child branch along with the initial intersection point defined as:

$$S_{branch} = C_n + \left[T_A - \left(\overline{C_n - O}\right).\vec{r}\right].\overrightarrow{r_{branch}} \tag{5.4}$$

where,

$\overrightarrow{r_{branch}}$, direction of intersection plane normal of branch ΔA

T_A, maximum value of projection of boundary points of A_{n+1} on associated new intersection normal

O, origin of workpiece co-ordinate system

Evaluation of successive cross-sections will be continued for the current branch until case 1 is reached. The algorithm repeats itself for every child branch detected, with the evaluated plane intersection point (S_{branch}) and changed plane direction ($\overrightarrow{r_{branch}}$) of respective branches. Every branch stores information of the indices of the parent branch and the number of child branches. Figure 5.13 represents the working of the algorithm schematically.

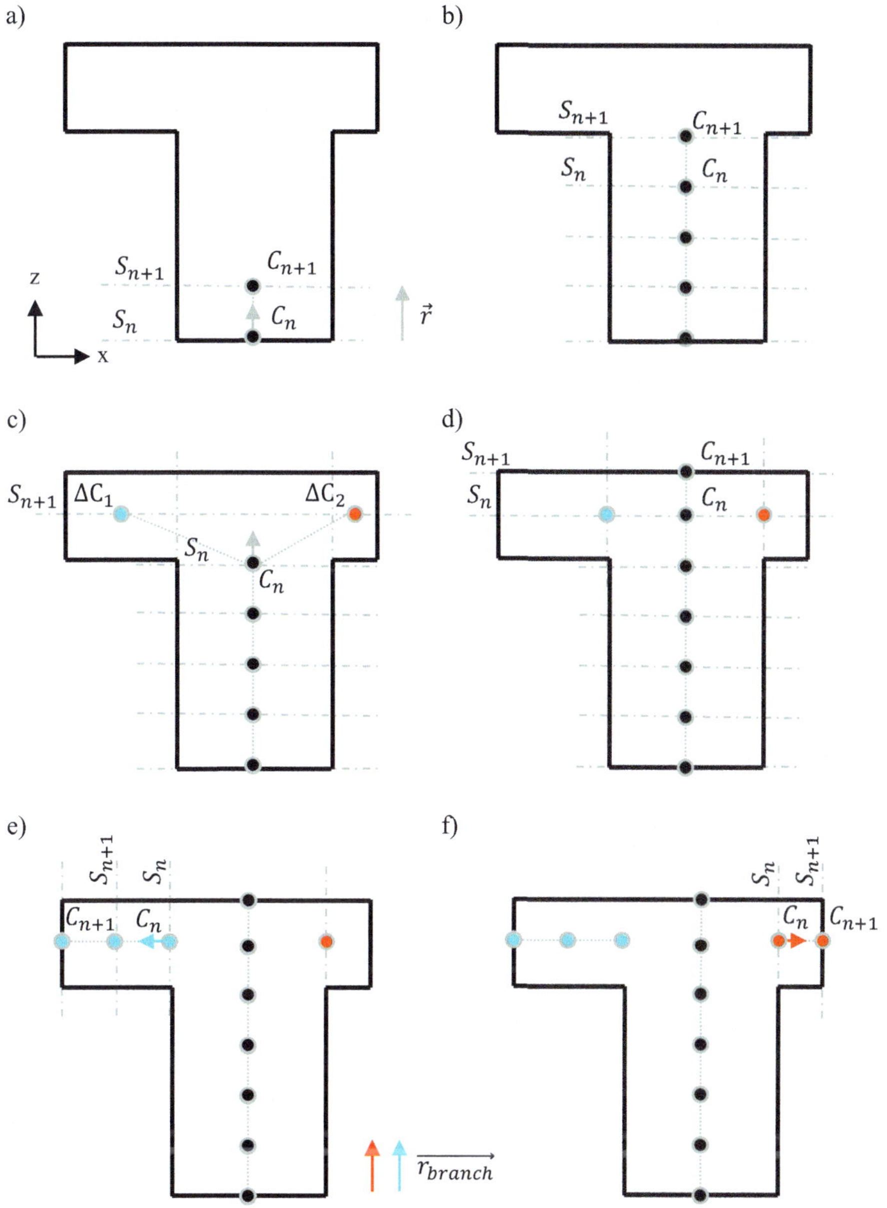

Figure 5.13 Schematic representation of procedural centroidal axis extraction of a parent and child branches of a STL-file format-based demonstrator geometry with two overhanging features

Successful determination of centroidal axis is followed by branch-wise mesh decomposition of the entire component using centroidal axis information and the parent-child depth tree information. Parting planes are generated with the initial centroid point and branch direction of every child branch. Mesh of the entire component is divided into two parts, while the facets below the plane belong to parent branch, the facets above and on the parting plane belong to the child branch. This mesh also consist of meshes of all other child branches. Mesh representing the current branch is segregated through ray intersection method of initial centroid of current branch onto the meshes. Parting of STL files result in hole generation leaving a closed loop of edges, which is then stitched with additional facets following the hole-filling algorithm mentioned by Ding et al. [65, 66].

Branch-wise water-tight meshes are then sliced further according to the user-defined slicing thickness to result in 2D-slice contours. For planar components, the slicing follows one of the principal axes along which the feature is protruding.

5.3.2. Pseudocode

Function: Centroidal axis extraction algorithm

Input: STL file, step_value, branch_direction [0,0,1], start_points [0,0, Z_{min} + 0.001]

Result: Returns arrays of centroids, links, start_points, branch_direction

for i =1 to length(start_points)

  if ≠1

    find base cross-section of current branch and remove additional starting points in start_points except the base cross-section and assign new branch direction

  end

  create **Struct** to store sliced polygon, branch number, direction, intersection_plane and centroid

  call **centroid function** to **return** values to **Struct**

    while flag == 1

      compute centroid of first/ base intersection plane (A_n) of current branch

      obtain cross-section (A_{n+1}) with step_value

      compute $A_{n+1} - (A_{n+1} \cap A_n)$

        if A_{n+1} , **then** flag == 1 and exit **centroid function**

        elseif Φ , **then** compute centroid for A_u and **return** value to centroids

        else

          for every $\Delta A_1, \Delta A_2, \Delta A_3, \ldots, \Delta A_n$ exceeding A_o, find new starting point based on $S_{branch} = C_n + \left[T - \left(\overrightarrow{C_n - O}\right).\vec{r}\right].\overrightarrow{r_{branch}}$ and associated direction based on direction change condition and append in start_points and branch_direction array

          save index of parent branch into every start_points

          Compute centroid for common area extracted $(A_u \cap A_l))$ and return value to centroids

  compute links L_{ij} from centroids array and append to Links array

  display centroidal axis with Centroids and Links array

end for loop

Function: Volume decomposition algorithm to extract sub-meshes of child branches

Input: struct centroidal axis, part mesh

Result: Returns decomposed meshes

for i =1 to length(start_points)

extract child branch and its corresponding centroidal axis

generate parting plane at first centroid of centroidal axis with the branch direction vector

divide part_mesh into below_mesh and above_mesh. above_mesh also includes facets interesting the parting plane.

segregate mesh representing the current child branch

generate branch_mesh with segregated above_mesh and by closing hole created during parting

end

Function: Slicing algorithm

Input: Branch meshes, slicing thickness, centroidal axis, branch direction vector

Result: Returns 2D slice contours

for i =1 to length (Mesh_parts)

for every element of start_points extract child branches and its corresponding centroidal axis

create parting plane at first centroid of centroidal axis with the branch direction vector

facets are divided into below_mesh and above_mesh. above_mesh also includes factes interesting the parting plane. above_mesh might also consist of meshes of other branches. The correct mesh representing the current branch is selected based on checking whether centroids belonging to current branch is contained in that mesh using ray intersection method.

end

5.3.3. Implementation Results

The developed algorithm has been implemented in MATLAB R2018a and tested on several test geometries and also on few demonstrator geometries. Figure 5.14 shows the implementation result of the developed algorithm on a STL-file format-based demonstrator geometry. Figure 5.14.b and c show the extracted centroidal axis along the initial +z direction, followed by branch separation along the features in the -y and +x axis. Figure 5.14.d shows the result of slicing algorithm on individual volume segments following the direction of respective centroidal axis.

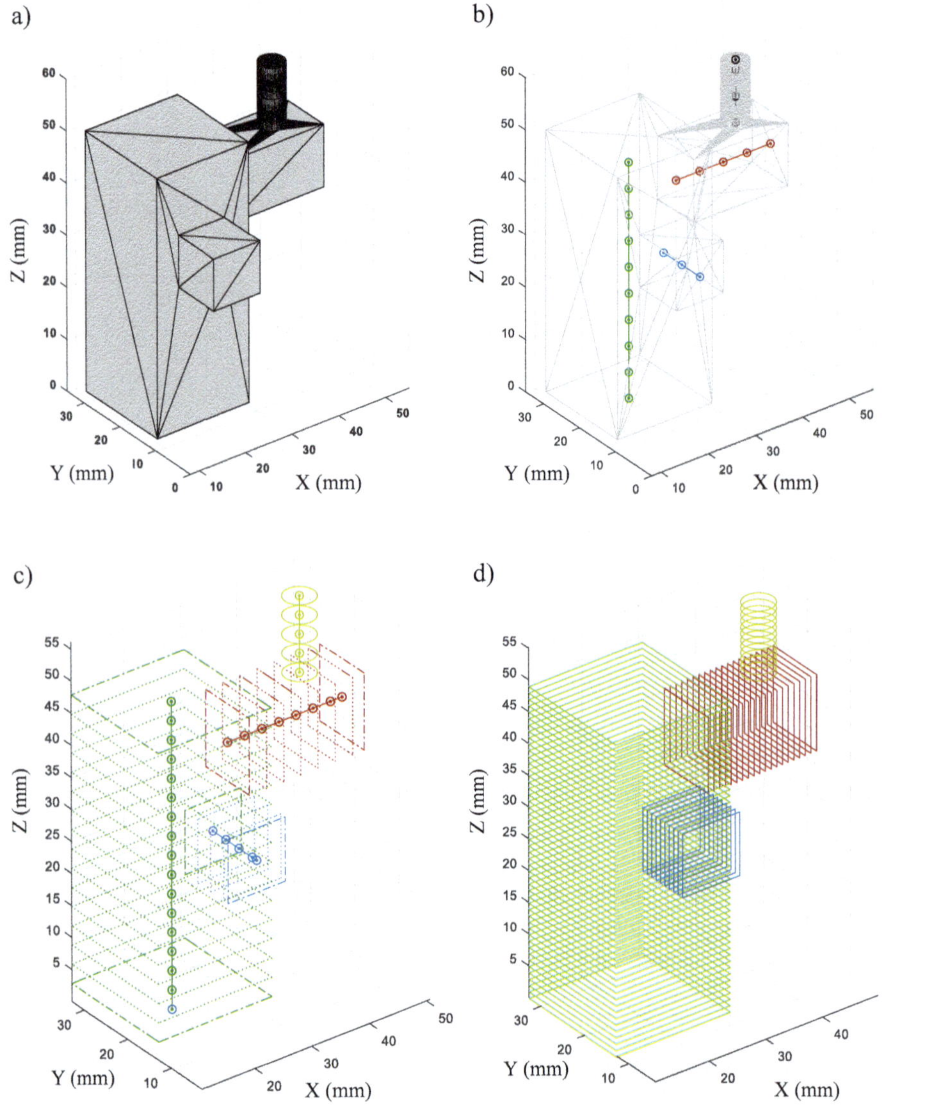

Figure 5.14 a) MATLAB representation of demonstrator component with overhanging features in STL-file format; b) Extracted centroidal axes of the demonstrator component; c) Volume segmentation based on extracted centroidal axis and determination of slicing axis; d) Planar slicing of segmented volumes along the direction of corresponding centroidal axis of the segment

Figure 5.15.a shows the hierarchical decomposition of demonstrator component into segments based on the computed centroidal axis. Figure 5.15.b shows the corresponding parent-child model tree as stored in the memory.

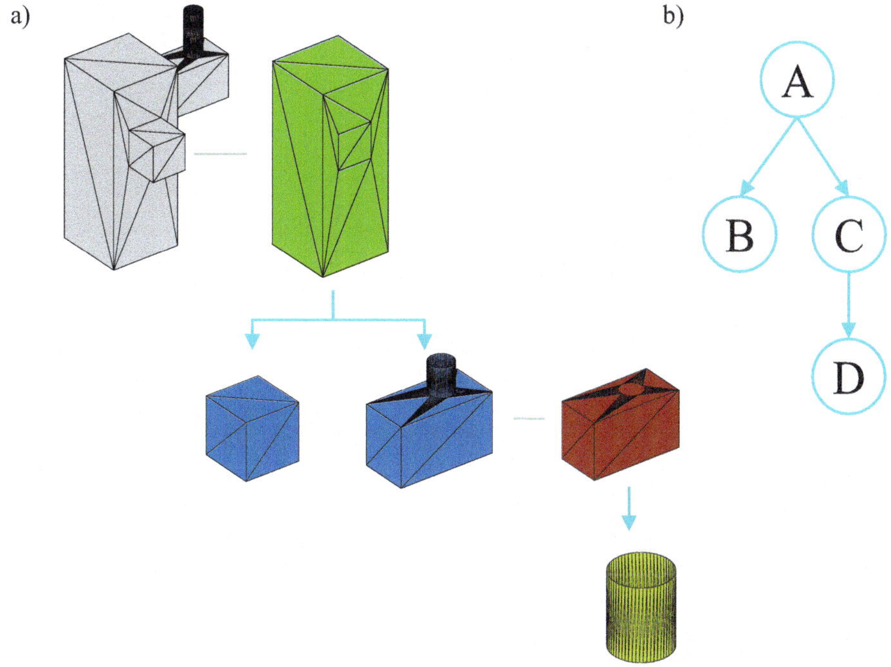

Figure 5.15 a) Hierarchical decomposition of demonstrator component into segments; b) Corresponding parent-child model stored in memory

Figure 5.16 and Figure 5.17 show the implementation of centroidal axis extraction, followed by volume segmentation and multi-direction slicing on demonstrators taken from automobile and railway sector respectively.

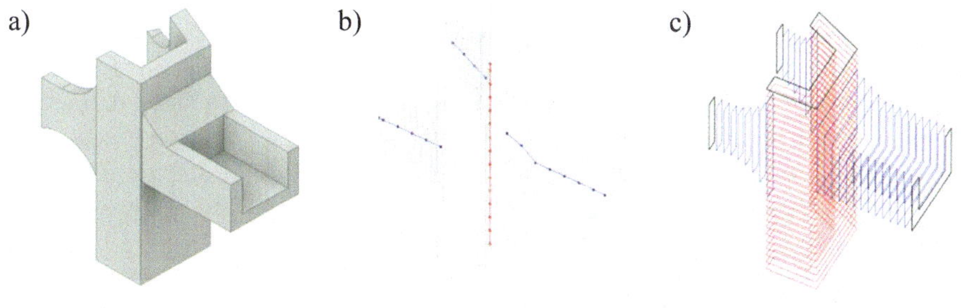

Figure 5.16 Implementation of volume segmentation and multi-direction slicing algorithm on an automobile demonstrator component: a) CAD Model, b) Tessellated mesh with the extracted centroidal axis enabling volume segmentation, c) Planar slicing of segmented volumes along the corresponding centroidal axis (not to scale)

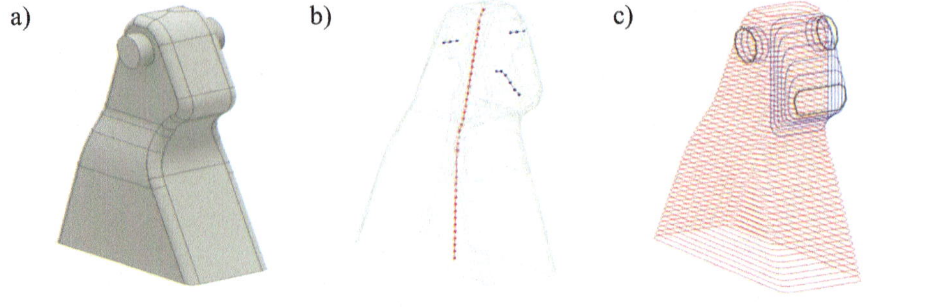

Figure 5.17 Implementation of volume segmentation and multi-direction slicing algorithm on a demonstrator component from railway sector: a) CAD Model, b) Tessellated mesh with the extracted centroidal axis enabling volume segmentation, c) Planar slicing of segmented volumes along the corresponding centroidal axis (not to scale)

5.3.4. Gradually-inclined Slicing

Gradually-inclined slicing is a special case of the volume segmentation algorithm, where the new segment is not an overhanging feature of the parent model but rather a gradually inclined feature that hinders deposition of additive material in a single fixed vector. Figure 5.18.a shows a demonstrator component chosen to test the algorithm. The first step is the determination of the centroidal axis along the principal axis (+z). The extracted centroidal axis along with the intersection plane is shown in Figure 5.18.b. It can be noticed from the side view in Figure 5.18.c that the centroidal axis changes its direction along the chosen principal axis. This inclination is determined by the angle made between successive links, L_{ij}. Links are connectors between any two Centroids $C_i$and C_j and are derived by

$$L_{i,j} = C_j - C_i, \tag{5.5}$$

where,

$$i = 1,..,n \; and \; j = i - 1$$

Slicing between two centroids C_n and C_{n+1} is carried out with the user-defined slice thickness and with the directional vector of the first link $L_{n,\ n+1}$. For the consecutive segment between centroids C_{n+1} and C_{n+2}, the slicing plane follows the directional vector of $L_{n+1,\ n+2}$. This results in a gradually inclined normal plane for slicing. However, setting centroidal axis extraction step close to or equal to the slicing thickness will result in adaptive layer-wise change in normal direction over all built layers. Figure 5.18.d shows the planar slicing of the component along the centroidal axis inclination.

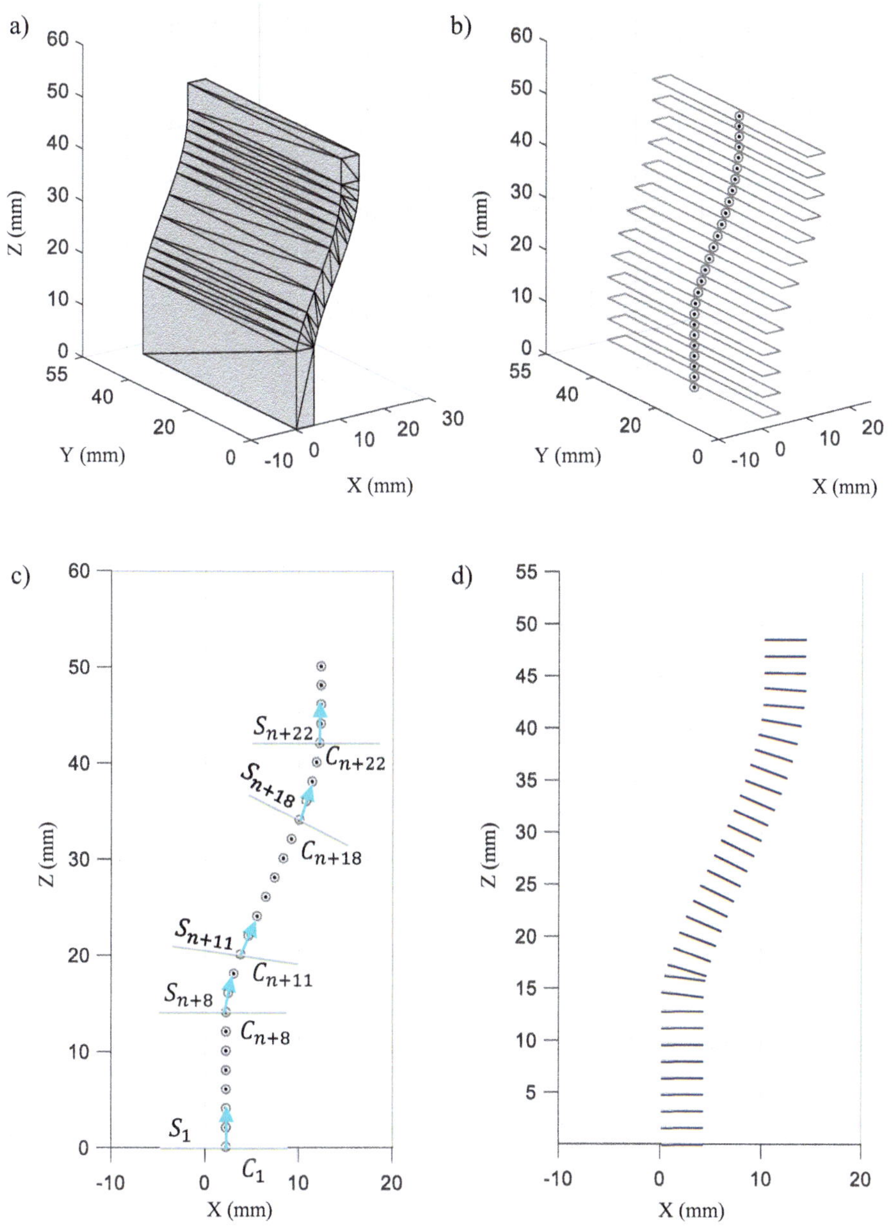

Figure 5.18 a) MATLAB representation of the gradually inclined demonstrator geometry; b) Extraction of centroidal axis along the principal axis (+z); c) Estimation of centroidal axis inclination determined by the angle made between successive links; d) Planar slicing of the component along the centroidal axis inclination

Figure 5.19 differentiates planar slicing along the principal axis (a) and planar slicing along the centroidal axis inclination (b) of the same demonstrator component. Figure 5.20 shows the implementation results on another similar demonstrator geometry.

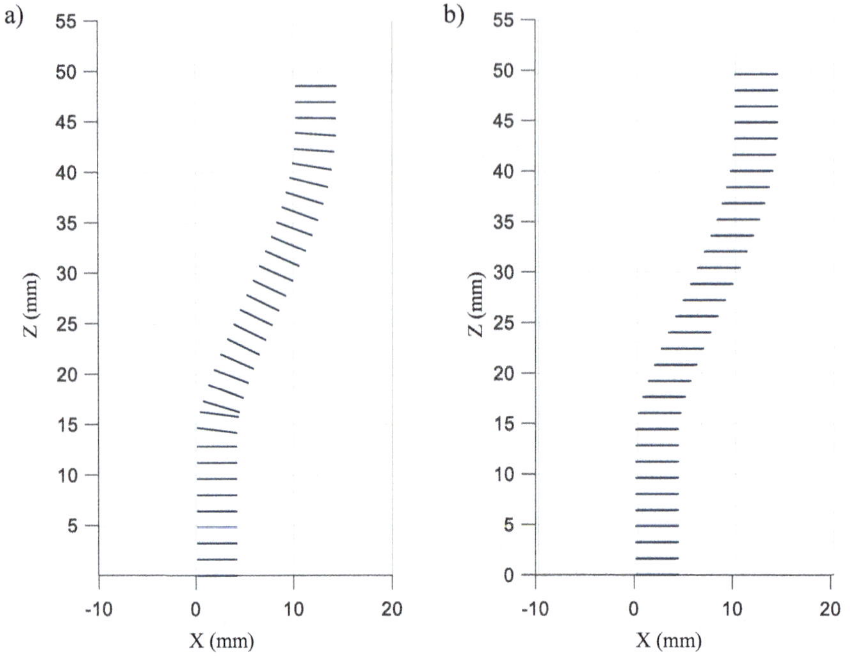

Figure 5.19 a) Planar slicing of the demonstrator component along the directional change of the centroidal axis; b) Planar slicing of the demonstrator component along the principal axis

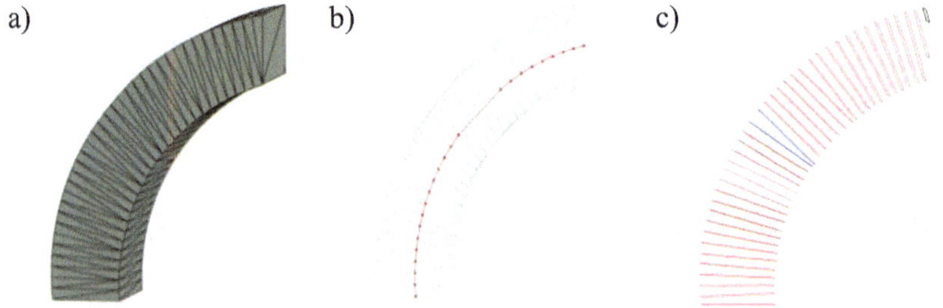

Figure 5.20 Implementation of gradually inclined slicing algorithm on a demonstrator component: a) CAD Model, b) Tessellated mesh with the extracted centroidal axis and determination of centroidal axis inclination, c) Planar slicing along the direction of the centroidal axis (not to scale)

5.4. NURBS-based Freeform Slicing

The kinematics of 6-axis industrial robots used commonly in DED processes enable fabrication of complex freeform geometries and hybrid structures, where the substrate could be a complex 3-dimensional substrate. Non-Uniform Rational B-Splines (NURBS) offer precise mathematical representation for both analytical geometric objects like simple lines, circles and arcs as well as for complex non-planar freeform curves and surfaces. Flexibility achieved through manipulation of NURBS control points and weights makes it ideal for computational modeling and toolpath generation of complex freeform surfaces. Calculations with NURBS directly can be done rather quickly and are numerically stable in comparison to working with the tessellated form (STL-file export) of freeform surfaces where all the mathematical representation of curves and surfaces are lost. Before describing the toolpath generation algorithms for NURBS surfaces, a brief review of Bézier and B-spline curves and surfaces are presented which are the building blocks of NURBS surfaces.

5.4.1. Bézier Curve and B-splines

Bézier Curves are binomial polynomial functions of parameter 't' ranging between 0 and 1, with coefficients taking the 'binomial' form as described below:

$$linear = (1-t) + t$$

$$quadratic = (1-t)^2 + 2.(1-t).t + t^2$$

$$cubic = (1-t)^3 + 3.(1-t)^2.t + 3.(1-t).t^2 + t^3$$

A generic function of Bézier curve can be described as:

$$\text{Bézier (n, t)} = \sum_{i=0}^{n} \binom{n}{i}.(1-t)^{n-i}.t^i.w_i \quad (5.6)$$

where, w_i is the weight given to the equation, which in turn are the co-ordinate values of the control points.

To gain further control over the curve property, they are rationalized, known as Rational Bézier curves. Ratio values provided to every control point dictate their extent of influence on the shape of the curve.

$$\text{Rational Bézier (n, t)} = \frac{\sum_{i=0}^{n} \binom{n}{i}.(1-t)^{n-i}.t^i.w_i.ratio_i}{\sum_{i=0}^{n} \binom{n}{i}.(1-t)^{n-i}.t^i.ratio_i} \quad (5.7)$$

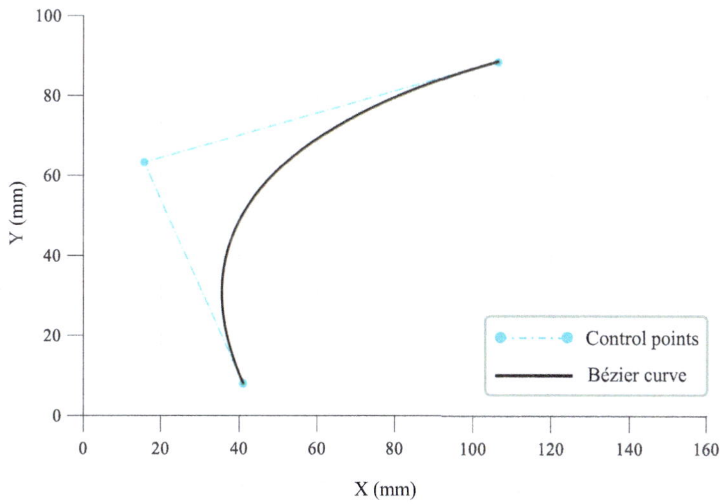

Figure 5.21 A quadratic Bézier curve with three control points

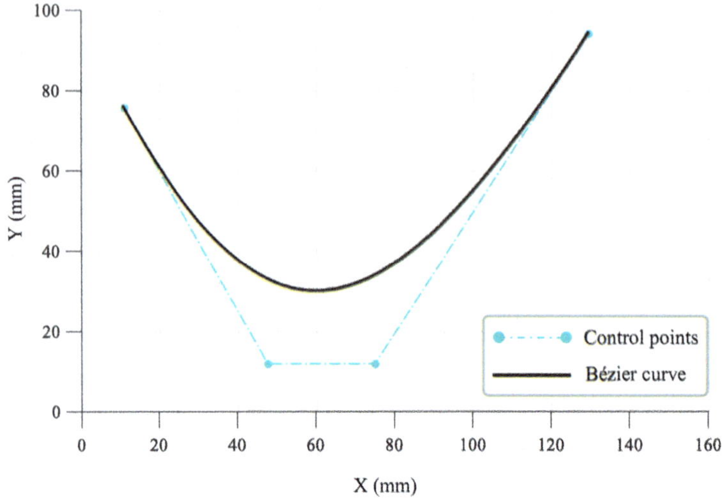

Figure 5.22 A cubic Bézier curve passing through four control points

Figure 5.21 shows a quadratic curve of degree 2 passing through 3 data points and Figure 5.22 shows a cubic Bézier curve of degree 3 passing through 4 control points. To generalize, a Bézier curve to pass through 'n' data points requires (n-1) degrees. To describe complex curves or shapes, a larger number of constraints are required along with a larger number of data points which then leads to a higher degree polynomial. Such a higher degree polynomial curve tends to be inefficient to process and is stated to be numerically unstable [144].

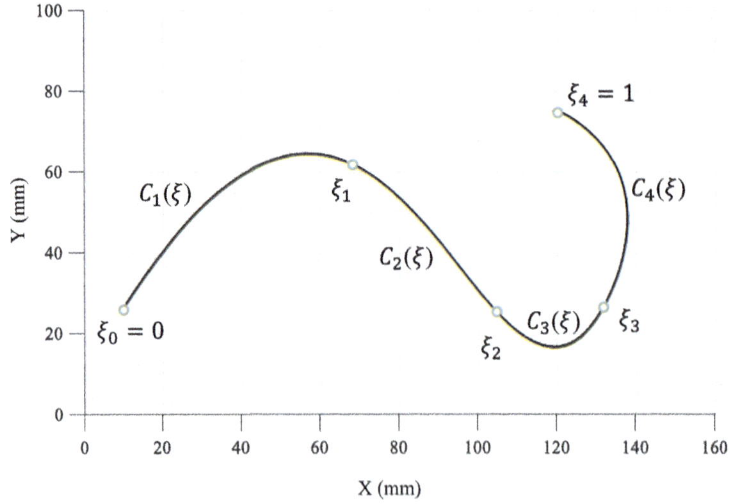

Figure 5.23 An example piecewise polynomial curve

The solution to this complexity is the use of B-splines, which are piecewise, polynomial interpolation curves, generated through polynomial interpolation over a set of points. To differentiate B-spline curve with Bézier curve, Bézier curves are defined as an interpolation of points, whereas B-splines are defined actually as an interpolation of curves.

Figure 5.23 shows a polynomial curve $C(\xi)$, consisting of four 3rd degree polynomial segments. $C(\xi)$ is defined on $\xi \in [0,1]$. The breakpoints ξ_1, ξ_2, ξ_3 *and* ξ_4 map into endpoints of the four polynomial segments. Figure 5.24 shows the same curve represented in cubic Bézier form, with 16 data points. Figure 5.25 represents a B-spline curve of degree 3 with the seven control points at {(5,10), (15,25), (30,30), (45,5), (55,5), (60,25), (50,30)}.

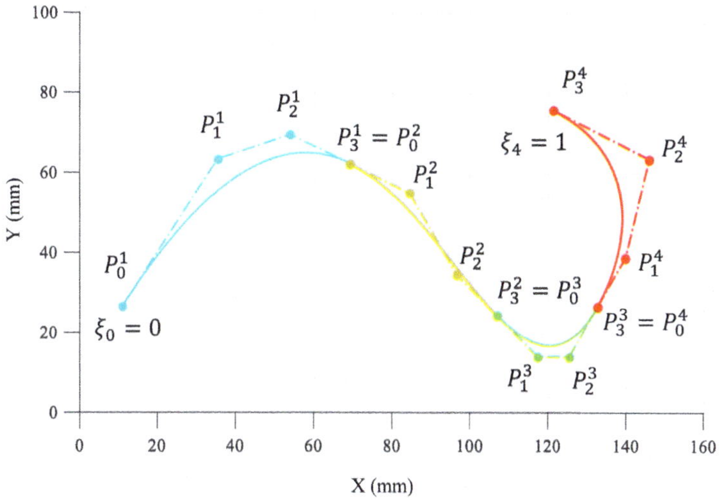

Figure 5.24 The B-spline polynomial curve of Figure 5.23 represented in Bézier form

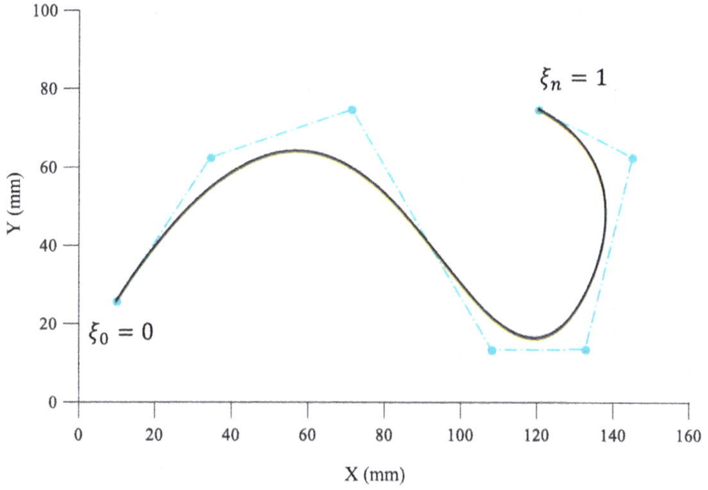

Figure 5.25 Representation of a B-spline curve of degree 3 with seven control points

The evaluation function of a generic B-spline of degree d and order $k = d + 1$, with n control points P_1 through P_n can be written with the parameter value ξ as:

$$C(\xi) = \sum_{i=1}^{n} P_i . N_{i,d}(\xi) \tag{5.8}$$

$$N_{i,0}(\xi) = \begin{cases} 1 & if\ \xi_i \leq \xi < \xi_{i+1} \\ 0 & otherwise \end{cases} \tag{5.9}$$

$$N_{i,d}(\xi) = \frac{\xi - \xi_i}{\xi_{i+d} - \xi_i} N_{i,d-1}(\xi) + \frac{\xi_{i+d} - \xi}{\xi_{i+d+1} - \xi_{i+1}} N_{i+1,d-1}(\xi) \tag{5.10}$$

The parameter k represents the knot interval where the section of the curve is defined. A B-spline curve with a degree (d) and n control points has the knot vector of length $d + n + 1$.

$$U = \{a_1, \ldots, a_{d+1}, u_{d+1}, \ldots, u_n, b_1, \ldots, b_{d+1}\} \tag{5.11}$$

The values in the knot vector are non-decreasing. The intervals of the knot vector (relative interval) are used for interpolations rather than the actual value in the polynomial function. If the knot interval is distributed uniformly over the curve interval, then the resulting shape is a uniform B-spline. Non-uniform knot interval results in Non-uniform B-spline. Collapsing knot intervals at the start and end of the curve results in a curve whose end controls points are also end points of the curve. i.e. when

$$a = a_1 = \cdots = a_{d+1}$$

$$b = b_1 = \cdots = b_{d+1}$$

By introducing weights to the control points, B-splines can be rationalized to form Rational B-splines. Combining non-uniformity with rationalization results in the formation of Non-Uniform Rational B-Spline (NURBS), a mathematical model that is the industry de-facto for representation of geometries and models in modern CAD systems [145].

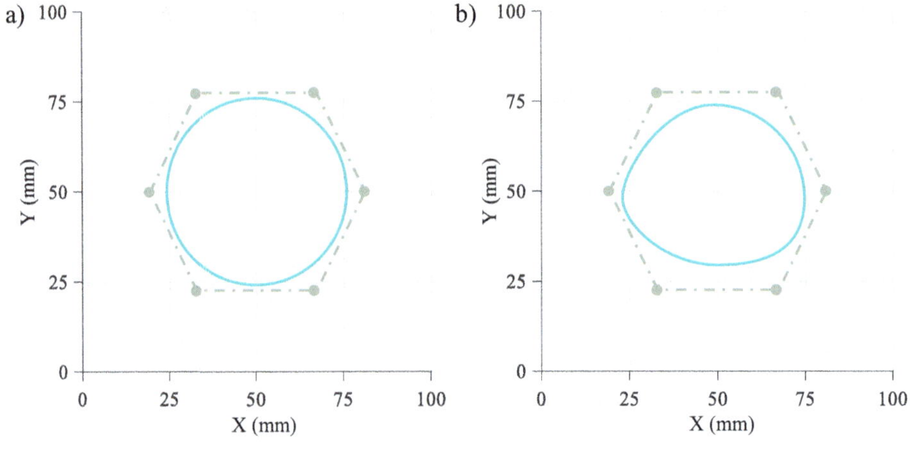

Figure 5.26 a) A rational uniform B-spline with equal weights; b) A rational uniform B-spline with unequal weights

NURBS curves and surfaces are the building blocks of geometric modeling, that can describe both regular and freeform geometries. Figure 5.26. shows two NURBS curve with same control points and degree, that represents a regular geometry (circle) and an irregular freeform geometry, by just adding different weights to the control points. The derivative of the basis function can be evaluated to compute tangents, normal and bi-normals of the B-Spline.

5.4.2. NURBS Surface und Volume

B-spline curves can be extended to surfaces by taking a bi-directional set of control points, two knot vectors and the product of the two B-spline curves [144]. B-spline based surfaces can be defined similarly with the two parameter values ξ and η, and their respective degrees p and q as:

$$S(\xi, \eta) = \sum_{i=1}^{n} \sum_{j=1}^{m} P_{ij} . N_{i,p}(\xi) . M_{j,q}(\eta) \tag{5.12}$$

Figure **5.27** shows a B-spline surface with the following parameters:

surf.knotvector(ξ)	= [0.0, 0.0, 0.0, 0.0, 1.0, 2.0, 3.0, 3.0, 3.0, 3.0]
surf.knotvector(η)	= [0.0, 0.0, 0.0, 0.0, 1.0, 2.0, 3.0, 3.0, 3.0, 3.0]
surf.degree(ξ)	= 3
surf.degree(η)	= 3
Number of control points	= 30

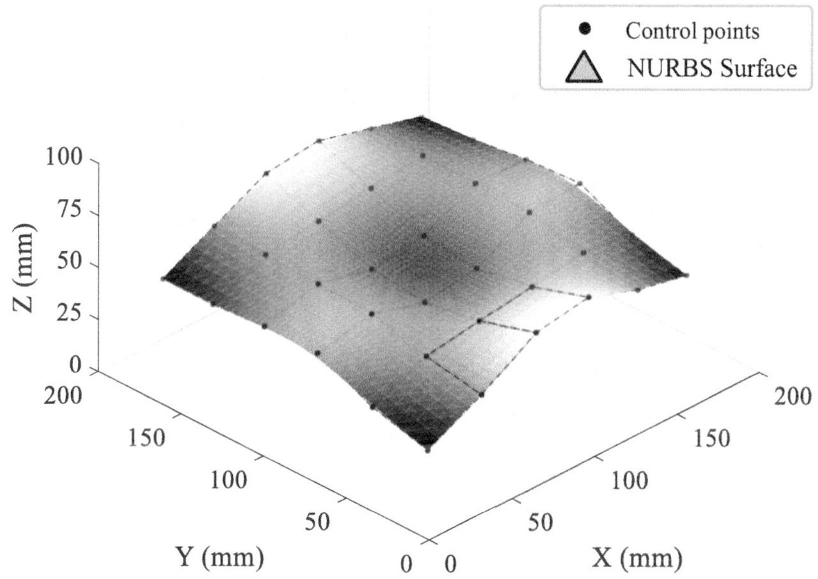

Figure 5.27 A freeform B-spline surface represented in the world co-ordinate system

Deposition of a freeform geometry is essentially a layer-wise deposition of several freeform surfaces that could be defined as NURBS surfaces. The deposition surface described in world coordinate system (X, Y and Z) (see Figure 5.28) can be mapped to a NURBS surface described in their parametric domain (ξ and η, represented alternatively as u and v) between 0 and 1, as shown in Figure 5.29. Any point in the NURBS surface can be mapped to real world coordinates using the equation (5.12) with the parametric domain. Figure 5.28 and Figure 5.29 shows a point both in its world co-ordinate domain and in the parametric domain.

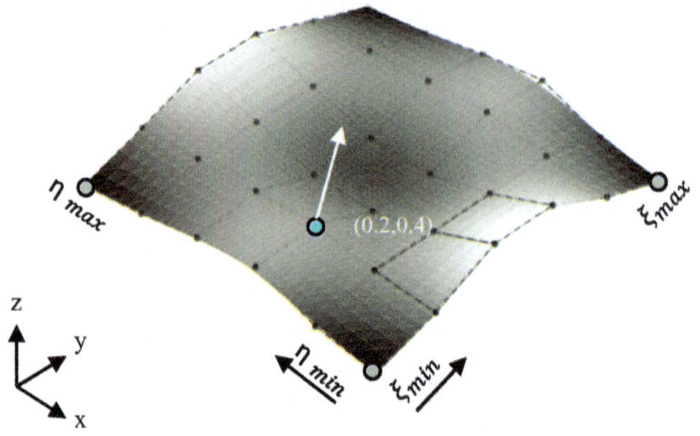

Figure 5.28 NURBS surface in the three-dimensional world co-ordinate system

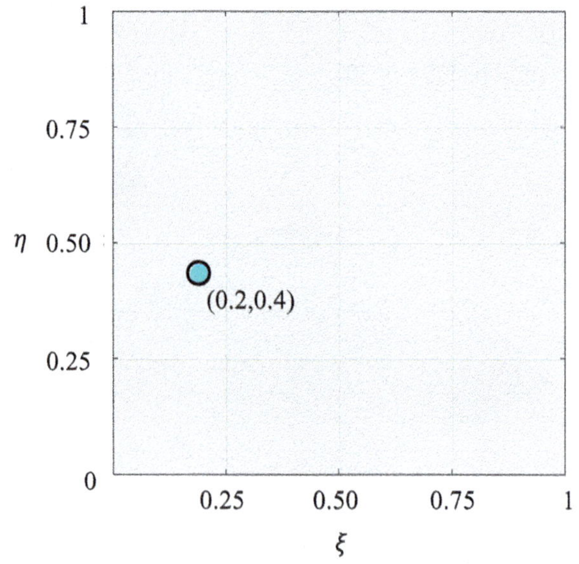

Figure 5.29 NURBS surface in the two-dimensional parametric domain ξ and η

Toolpaths along 'y' direction are generated by extracting splines along parametric 'η' direction at defined intervals of perpendicular parametric direction, $\Delta\xi$. Similarly, toolpaths along 'x'' direction can be generated by extracting splines along parametric 'ξ'' direction with $\Delta\eta$ intervals. Such toolpaths that are evaluated at regular intervals in the parametric domain, are known as iso-parametric splines. These, however, may or may not be equi-distant in the world coordinates. For regular planar surfaces and symmetrical surfaces, they would be equi-distant in the world coordinates. For all other freeform surfaces, a correlation between the world coordinate system and the parametric domain is required to generate equi-distant toolpaths. Equi-distant toolpaths are necessary for a defect-free and crater-free deposition process and to achieve the required layer-wise geometrical accuracy which is one of the deciding factors for build stability. The following sections describe the development of toolpath generation algorithms for the above-mentioned cases of iso-parametric and equi-distant splines.

5.4.3. Volume slicing and Toolpath Generation

Toolpath generation of a freeform geometry begins with the slicing of the additive volume with a base surface. The procedure, algorithm and the OCCT functions are the same as the one discussed in section 5.2 for non-planar slicing of STEP-file format-based geometries. The chosen base surface or foundation face in this case will be a B-spline surface rather than a cylinder or a cone. The sliced surfaces after slicing will be B-rep surfaces which is then downcasted to B-spline surface using the *brep.surface()* OCCT method. This is followed by the extraction of degrees, knot vectors and control points of the surface.

Toolpath generation of a B-spline or NURBS surface is essentially the extraction of B-spline curves on the NURBS surface in the parametric domain. For this purpose, an additional python library that entails the geometric computation algorithms as methods for working with NURBS called 'geomdl' has been used [142]. The extracted B-spline geometry information from the OCCT method has been used to generate back the B-spline, but as a geomdl B-spline object. This has been used for further computations using the methods and classes of geomdl library.

5.4.4. Iso-parametric Toolpath

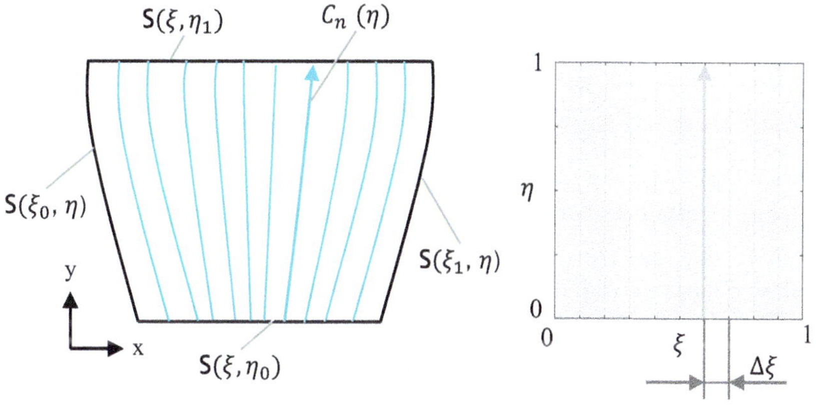

Figure 5.30 Iso-parametric NURBS curves along 'y' direction represented in world coordinates and parametric domain

Iso-parametric toolpaths along '*y*' direction is generated by extracting splines along parametric 'η' direction at defined intervals of perpendicular parametric direction, $\Delta\xi$. $\Delta\xi$ can be evaluated using the following two equations:

$$Sample\ size = \frac{x_{max} - x_{min}}{hatch\ distance} \tag{5.13}$$

$$\Delta\xi = \frac{1}{sample\ size} \tag{5.14}$$

The first spline is generated by evaluating surface points at ($\xi = 0, \eta_n$) in workpiece coordinate system, where n varies between 0 and 1. Lower the evaluation interval ($\delta\,\eta$) between 0 and 1, higher the number of spline points and smoother the curve, and so is the end-effector movement. The surface evaluation can be carried out through two of the geomdl methods, namely *surf.evaluate_single()* or through *operations.normal()*. The former method returns the coordinates, whereas the latter returns a list containing coordinates and the respective normal vector. The latter method has been used in this algorithm to extract surface points along with the unit normal vector that is necessary for orientation of the deposition head to the surface. These points are then stored in an array to from the first spline. Following splines are calculated in the similar manner at ($\xi + \Delta\xi, \eta_n$) until $\xi \rightarrow 1$. Figure 5.31 shows an iso-parametric curve C_n defined by a series of evaluation points $P_1^n, P_2^n, \ldots, P_n^n$ in the (x, y, z) domain and the corresponding representation in the parametric domain (ξ, η).

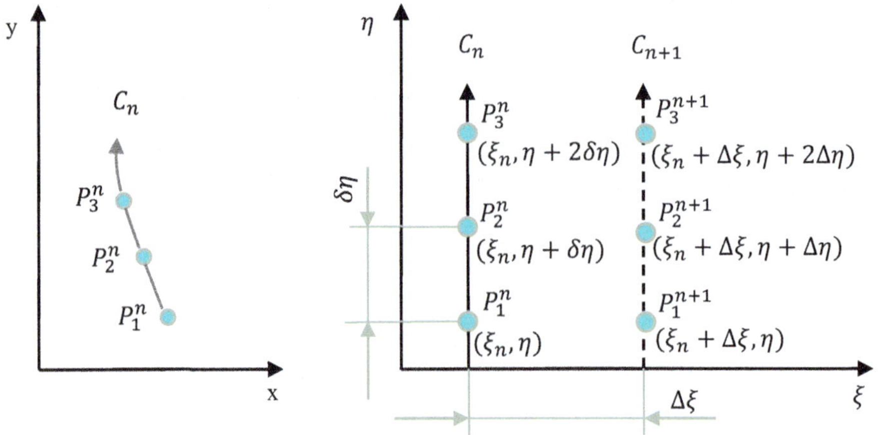

Figure 5.31 Definition and evaluation of iso-parametric B-spline curve in the world coordinate system and in the parametric domain respectively

Similarly, toolpaths along 'x'' direction can be generated by extracting splines along parametric 'ξ'direction with $\Delta\eta$ intervals calculated as:

$$Sample\ size = \frac{x_{max} - x_{min}}{hatch\ distance} \tag{5.15}$$

$$\Delta\eta = \frac{1}{sample\ size} \tag{5.16}$$

The first spline is then generated by evaluating surface points at ($\xi_n, \eta = 0$), where n varies between 0 and 1 at evaluation interval $\delta\,\xi$. Following splines are evaluated with parametric values at ($\xi_n, \eta + \Delta\eta$) until $\eta \rightarrow 1$ (as shown in Figure 5.32).

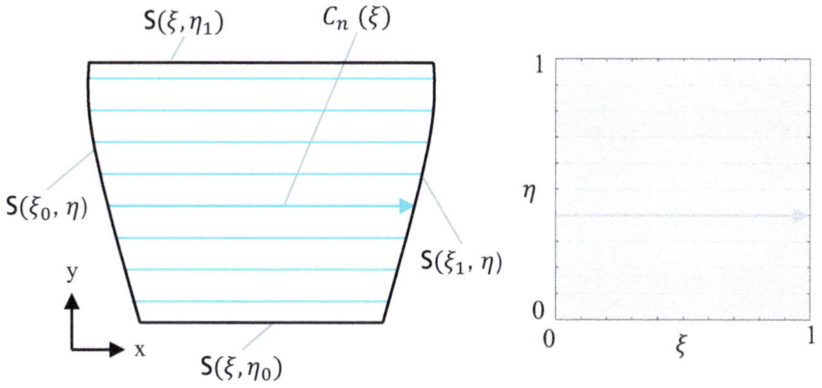

Figure 5.32 Iso-parametric NURBS curves along 'x' direction represented in world coordinates and parametric domain

5.4.4.1. Pseudocode

Function:	Iso-parametric toolpath generation
Input:	STEP-file, hatch_distance, hatch_direction
Result:	Returns list of splines with surface co-ordinates and unit normal vectors

Read step file using OCCT ReadFile() function

Extract B-spline surfaces with knots from STEP file by iterating through OCCT TopologyExplorer() function

Convert OCCT shape to BRep surface and downcast to B-Spline surface

Extract ξ degree, η degree, control points array, ξ knot vectors and η knot vectors from the extracted shape using OCCT functions

Generate geomdl B-spline surface using the extracted surface information

Evaluate $\Delta\xi$ and $\Delta\eta$ from sample size

if hatch_direction is parallel to X-direction

  while $\eta \leq 1.0$

    while $\xi \leq 1.0$

    extract point of normal at (ξ, η) using Geomdl operations.normal function and return to spline list

    $\xi += \delta\xi$

  Append to global spline list

  $\eta += \Delta\eta$

elseif hatch_direction is parallel to Y -direction

  while $\xi \leq 1.0$

    while $\eta \leq 1.0$

extract point of normal at (ξ, η) using Geomdl operations.normal function and return to spline list

$\eta += \delta\eta$

Append to global spline list

$\xi += \Delta\xi$

end

5.4.5. Equi-distant Toolpath

Even though the iso-parametric toolpaths are generated at equal intervals of parametric domain, these do not ensure or result in equi-distant B-spline curves or toolpaths in the workpiece coordinate system. Therefore, an alternate approach to determine the parametric interval for every adjacent spline point has been developed. By estimating the curve distance along the surface between the two adjacent spline points in the workpiece coordinate system evaluated for every small parametric increment, rather than a fixed interval ($\Delta\xi$ *or* $\Delta\eta$), the correct parameter (ξ, η) can be searched that results in equi-distant toolpaths. Figure 5.33 shows a graphical representation of such equi-distant B-spline curves (or toolpaths) along the direction of y-axis for the surface shown in Figure 5.30. The algorithm developed for evaluation of these equi-distant curves has been inspired and adopted from a similar approach used in 5-axis machining of freeform surfaces to generate iso-scallop toolpaths developed by Can and Ünüvar [146].

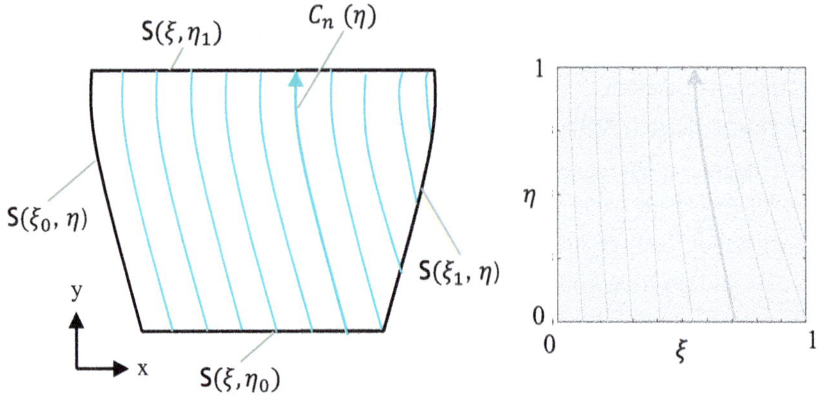

Figure 5.33 Equi-distant NURBS curves along 'y' direction represented in world coordinates and parametric domain

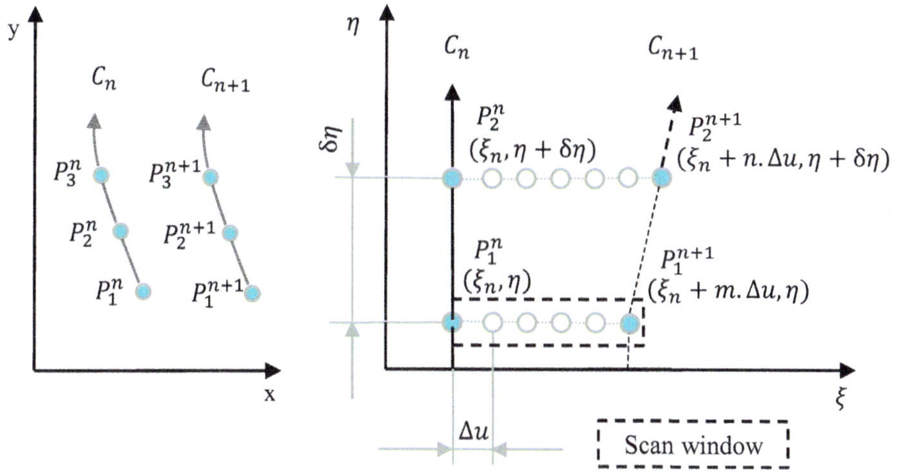

Figure 5.34 Definition and evaluation of equi-distant B-spline curve in the world coordinate system and in the parametric domain respectively

Equi-distant toolpath generation along 'η' direction begins with the extraction of first spline C_1 at $\xi = 0$, which is the boundary curve. The procedure is identical to the extraction of first spline of iso-parametric toolpaths. The second adjacent spline is the first equi-distant toolpath, C_2. To generalize, and as shown in Figure 5.34, let C_n be the first spline or an existing spline and C_{n+1} be the spline to be evaluated. Points on C_n, namely $P_1^n, P_2^n, \ldots, P_n^n$, are evaluated at an evaluation interval of δ η. The first equi-distant point P_1^{n+1} on C_{n+1} is searched from P_1^n point along the x- direction, i.e. the same η parametric value at a search interval of Δu (mentioned as 'scan window' in the Figure 5.34). The intermediate point P_i^{n+1} $(\xi_n + \Delta u, \eta)$ is evaluated and the distance between the two points P_1^n and P_i^{n+1} on the NURBS surface in the world coordinate system is determined. This search continues until the estimated distance reaches the user-defined hatch distance. This point then becomes the first point on the adjacent equi-distant toolpath $P_1^{n+1}(\xi_n + m.\Delta u, \eta)$. The second point on the curve C_{n+1} is scanned similarly from the second point P_2^n on the curve C_n along the parametric ξ direction until $P_2^{n+1}(\xi_n + n.\Delta u, \eta)$ is reached whose physical surface distance from P_2^n is equal to the hatch distance. The process is carried out iteratively until all the points on C_n has been computed with its equivalent C_{n+1} points or when η reaches 1. The curve C_{n+1} is now an equi-distant curve or toolpath. Following toolpaths are computed until ξ reaches 1. Δu is set to be $(1/10)^{th}$or $(1/100)^{th}$ of $\Delta\xi$, calculated for the case of iso-parametric. Smaller the Δu value, higher will the accuracy of the calculated toolpath. However, this also has an enormous effect on the computational time, especially when setting very low values like $(1/100)^{th}$ of $\Delta\xi$.

5.4.5.1. Pseudocode

Function: Equi-distant toolpath generation
Input: STEP file, hatch distance, hatch direction
Result: Returns list of splines with surface co-ordinates and unit normal vectors

Read step file using OCCT ReadFile() function
Extract B-spline surfaces with knots from STEP file by iterating through OCCT TopologyExplorer function
Convert OCCT shape to BRep surface and downcast to B-spline surface
Extract ξ degree, η degree, control points array, ξ knot vectors and η knot vectors from the extracted shape using OCCT functions
Generate geomdl B-spline surface using the extracted surface information
#Generation of first spline C_1
for η = 0 to 1
    return operations.normal(ξ = 0, η) to spline list
    η += δη
    Append to global spline list
end
#Generation of further splines $C_2, \ldots, C_n$
while $\xi \leq 1$
    for η = 0 to 1
        $\xi_n = \xi_{n-1} + \Delta u$
        Evaluate distance between (ξ_n, η) and (ξ_{n-1}, η)
        while distance < hatch_distance
            ξ_n += Δu
            Goto evaluate distance function again
        η += δη
    end
    Append the curve C_n to global spline list
end

5.4.5.2. Optimization algorithm

In the calculation of equi-distant toolpaths, the search interval Δu plays a key role in determining the accuracy of the toolpaths generated, but also decides the computational time required. It was mentioned earlier that Δu should be chosen $(1/10)^{th}$or $(1/100)^{th}$ of $\Delta\xi$ calculated for iso-parametric toolpaths. Choosing $(1/100)^{th}$ of $\Delta\xi$ can lead to roughly 100 search iterations for every individual surface point, thereby consuming longer computational time. To optimize this, an optimization algorithm has been developed that chooses the Δu dynamically during the run-time of the program.

Initially, Δu is assigned a constant pre-defined value until 70 – 80 % of the required hatch distance has been calculated. For further search iterations, Δu is decreased by $(1/10)^{th}$ that generates newer scan points much closer to the previous point. This increment is continued until around 90 % of the required hatch distance is arrived. For the last search iterations, Δu is decreased further by $(1/100)^{th}$ which leads to accurate determination of the adjacent toolpath point with a precision in 3 decimal places. This technique helps in evaluating equi-distant toolpaths that are precise and smooth, yet in an optimized way.

5.4.6. Implementation Results

Volume slicing, NURBS surface extraction, generation of iso-parametric and equi-distant toolpaths have been implemented in Python programming language along with some methods of the libraries pythonocc-core and geomdl. Figure 5.35 and Figure 5.38 shows the python representation of two skewed NURBS surfaces sliced from two demonstrator geometries.

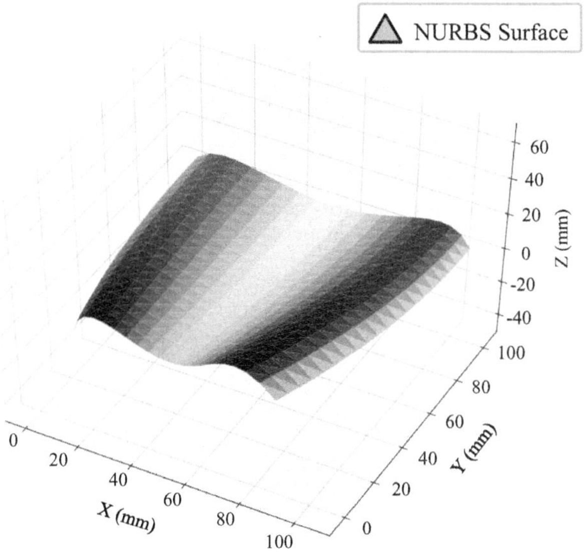

Figure 5.35 Representation of a NURBS surface in world coordinate system for demonstrator geometry 1

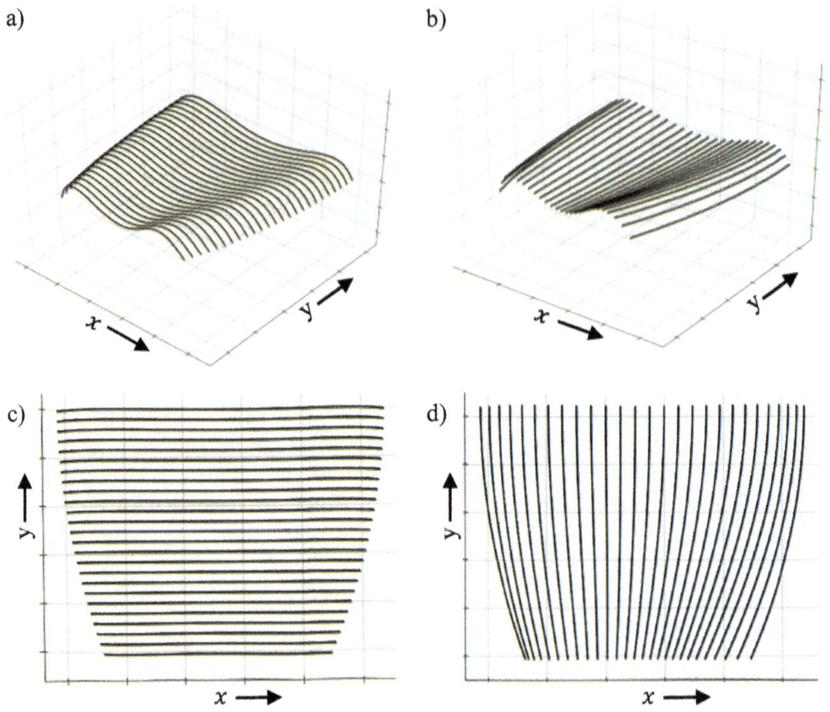

Figure 5.36 Extraction of iso-parametric toolpaths for demonstrator surface 1: (a) and (c) along x-direction; (b) and (d) along y-direction

Figure 5.36 shows the extracted iso-parametric toolpaths using the algorithm developed in section 5.4.4 along x- and y- directions. Top view of the representation (Figure 5.36.c and Figure 5.36.d) shows clearly that the toolpaths created in equal intervals of parametric domain ξ and η do not yield equally spaced splines in the physical space. Higher the skewness of the surface, higher is the inequality in the distance between toolpaths. This effect can be seen evidently in the toolpaths extracted along y direction.

Figure 5.37 shows equi-distant toolpaths extracted for the same demonstrator surface 1, along x and y directions. In contrast to the iso-parametric toolpaths, the equidistant toolpaths are seperated by the hatch distance in the physical domain at all points of the curve.

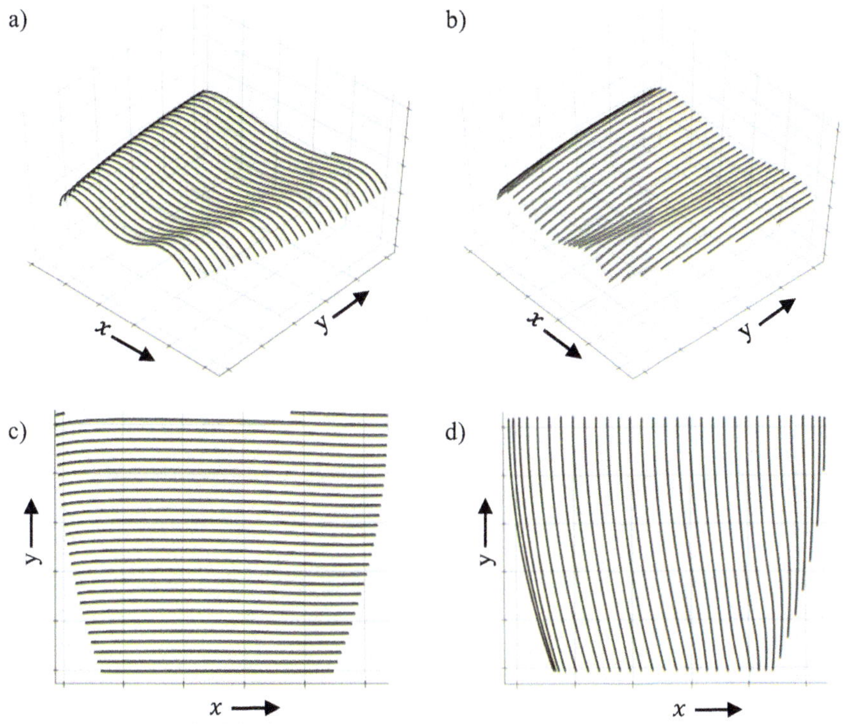

Figure 5.37 Extraction of equi-distant toolpaths for demonstrator surface 1: (a) and (c) along x-direction; (b) and (d) along y-direction

Similar to the demonstrator surface 1, Figure 5.39 and Figure 5.40 shows the iso-parametric and equi-distant toolpaths extracted for demonstrator surface 2 along x and y directions respectively.

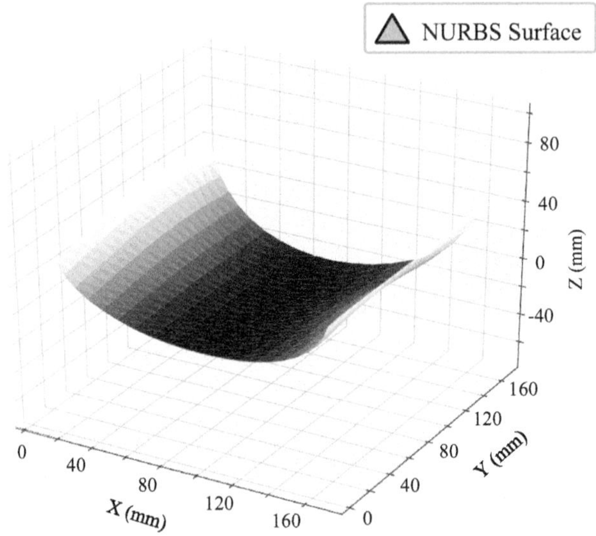

Figure 5.38 Representation of a NURBS surface in world coordinate system for demonstrator geometry 2

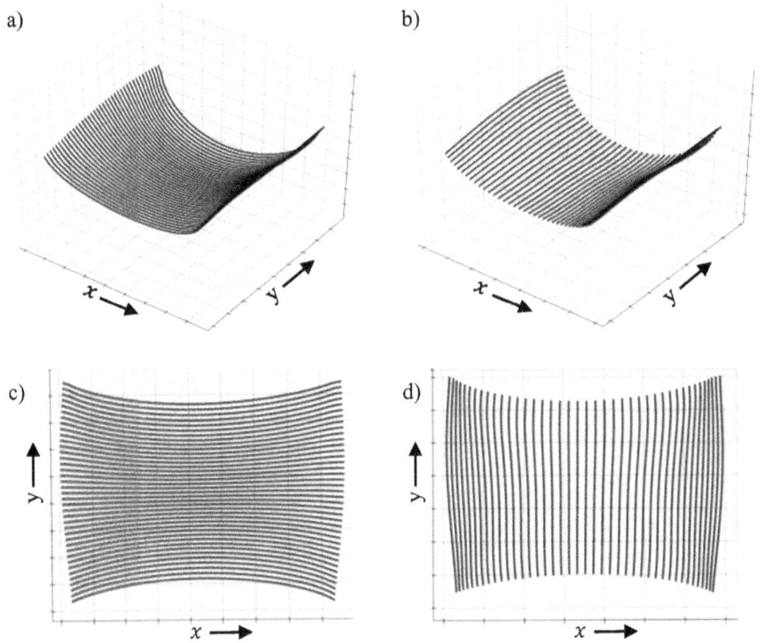

Figure 5.39 Extraction of iso-parametric toolpaths for demonstrator surface 2: (a) and (c) along x-direction; (b) and (d) along y-direction

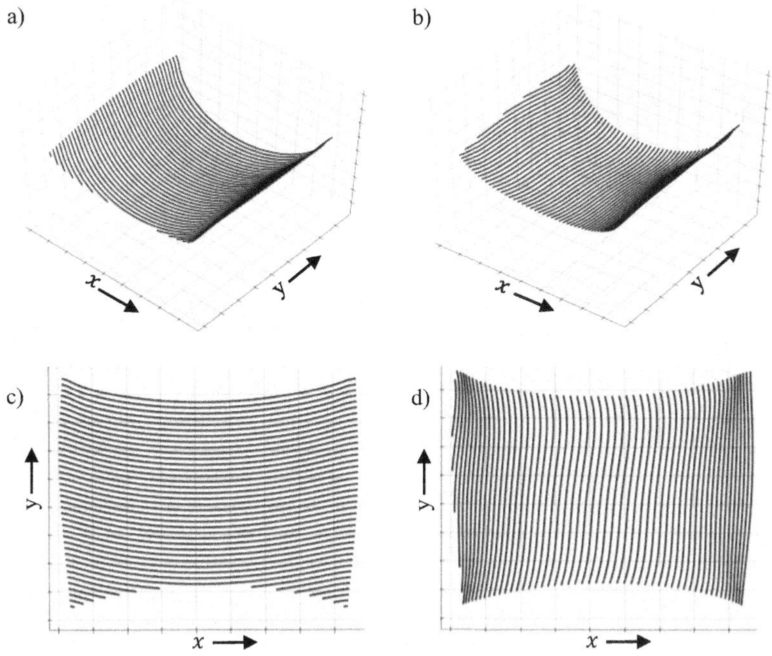

Figure 5.40 Extraction of equi-distant toolpaths for demonstrator surface 2: (a) and (c) along x-direction; (b) and (d) along y-direction

The developed algorithms have been tested on an industrial use-case scenario. Figure 5.41.a. shows an abstract 3-dimensional freeform geometry with non-uniform wall-thickness of the so-called metal leading edge from the aerospace sector. Metal leading edge acts as a protective metal cap for an aircraft turbofan engine blade. The represented hybrid component consists of a forged substrate with uniform wall-thickness upon which material can be deposited to achieve the desired cross-section. A CAD representation of a demonstrative freeform substrate and the to be deposited additive geometry or additive volume is shown and differentiated by color in Figure 5.41.a.

For toolpath generation, the additive geometry is initially sliced with the non-planar freeform base surface extracted from the substrate. The sliced freeform surface after slicing is visualized as a NURBS surface representation in Figure 5.42. For easier and better visualization only one half (left-half) of the complete NURBS surface is shown. Figure 5.43 shows the toolpaths generated for the NURBS surface along the x-direction (or parametric ξ-direction).

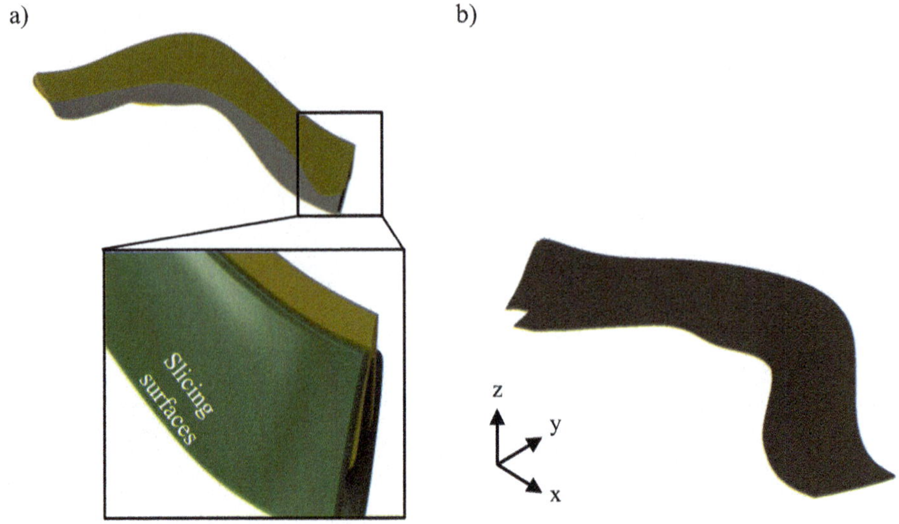

Figure 5.41 a) Volume slicing of additive geometry with non-planar freeform slicing surface; b) CAD representation of the demonstrative freeform sliced surface (not to scale)

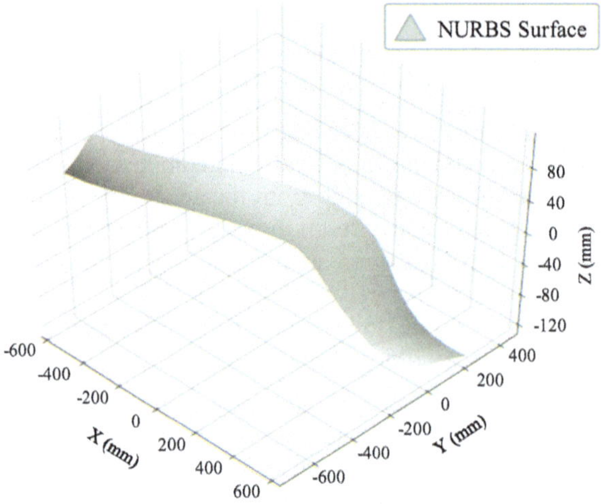

Figure 5.42 First slice (left-half) after slicing of the additive geometry with the freeform base surface represented as a NURBS surface

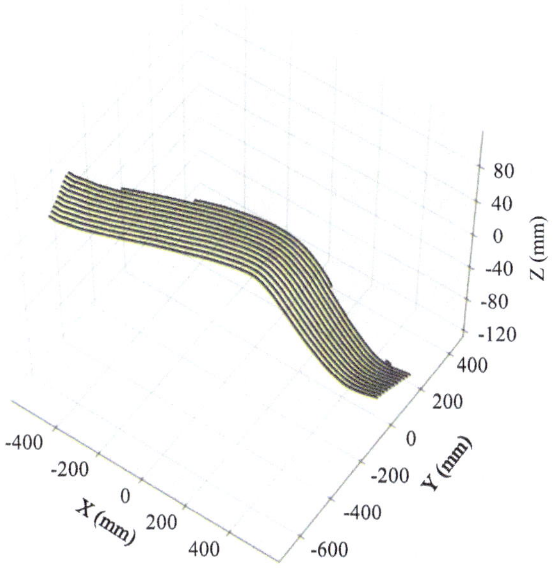

Figure 5.43 Generation of equi-distant toolpaths along the x-direction (or parametric ξ-direction) of the NURBS surface shown in Figure 5.42.

5.5. Segmentation and Toolpath Generation for Thin-Wall Components

Thin-wall components are characterized by a large base surface area with rib-like strengthening structures and pockets that makes the structure lighter yet stiffer [16]. Aerospace structural components are predominantly thin-wall structures that are manufactured from alloys of high strength and lightweight materials like Titanium and Aluminum. In such structures, the substrate plate can be designed to be a part of the final component making it a hybrid design.

Such integration leads to choosing a substrate plate that is only few millimeters thicker than the final component, which in-turn requires heavier clamping tools and bulkier base plate for heat dissipation. Toolpath generation for such thin wall geometries is another challenge that has been mentioned in the recent publications owing to its complexity and inadequate algorithms in the existing commercial software packages [147].

5.5.1. Contour-based Toolpath Generation

5.5.1.1. Contour-based offset toolpath generation

One of the most frequently used and commonly available toolpath pattern or hatching technique in additive manufacturing software packages is the contour-based offset toolpath generation function. In this, toolpaths within a layer are generated by creating several offsets of the outer and inner boundary towards the core or skeleton of the geoemtry, i.e. Outside-in. This technique is suitable for bulk volume components and also for thin-wall components with equal wall thickness. However, majority of the thin-wall components have variable wall-thickness over the geometry, thereby leading to non-equidistant toolpaths within the innermost layers. Such toolpaths lead to either formation of voids or can result in excessive weld bead overlap leading to localized higher builds. One such example has been depicted in Figure 5.45 for the demonstrator component shown in Figure 5.44.a. Figure 5.45 also shows three zoomed-in views of the toolpath sections where this effect is evidently visible. Using this toolpath pattern for high-deposition rate DED process can lead to generation of afore-mentioned voids, craters and binding defects or local material peaks eventually leading to process instability and part defect. Therefore, an alternate toolpath generation pattern is required that can result in equi-distant toolpaths.

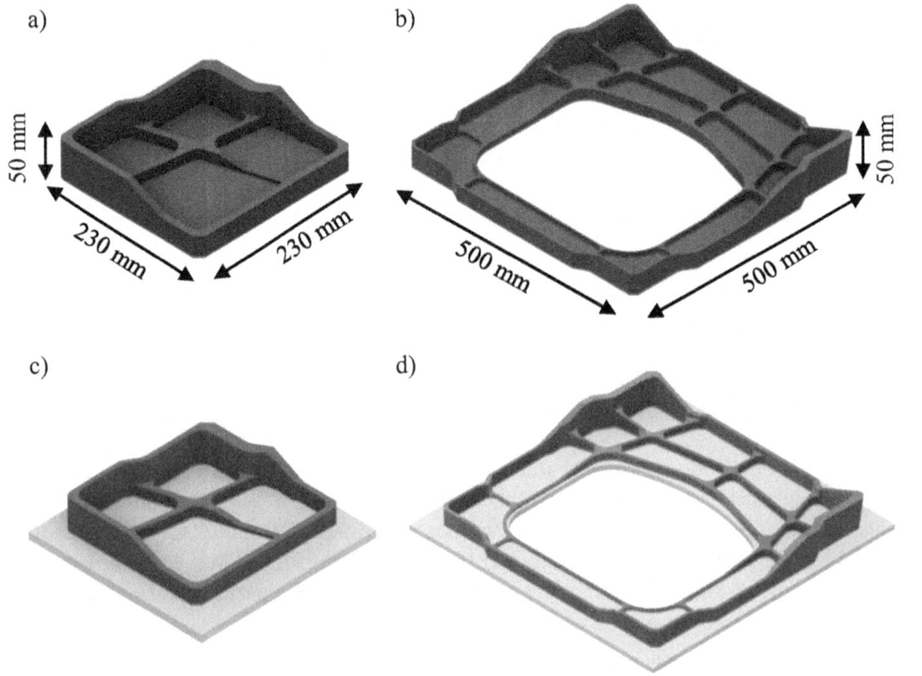

Figure 5.44 CAD representation of: a) demonstrator component 1, b) demonstrator component 2; Design modification for DED process with integration of base surface into substrate for: c) demonstrator 1, d) demonstrator 2

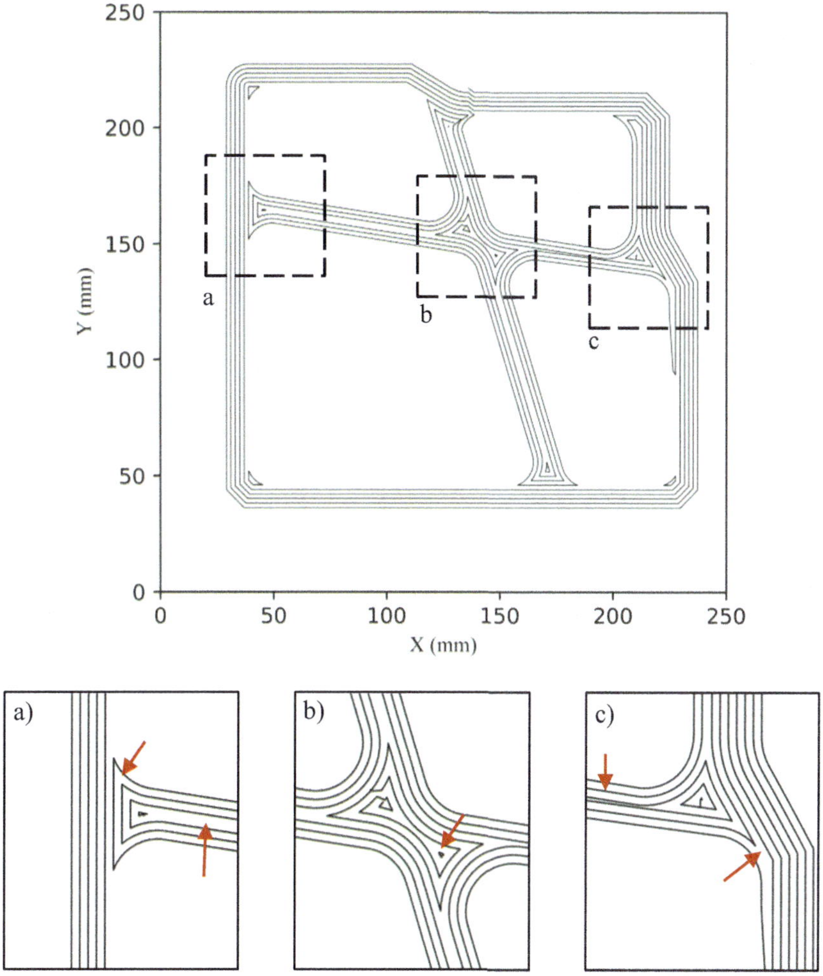

Figure 5.45 Contour-based toolpath generation from outside to inside for demonstrator geometry in Figure 5.44

5.5.1.2. Medial-axis based toolpath generation

To eliminate the inadequacies of the contour-offset based toolpaths which is essentially oustide-in, an inside-out offset approach has been proposed as early as 2001 by Kao and Prinz for simple DED geometries that result in void-free deposition toolpaths [148]. For this purspose, a mid-line or the so-called skeleton of the sliced layer was initally derived. Toolpaths were then generated by offseting this mid-line towards boundary, i.e. inside-out. This mid-line is the computed medial-axis of the sliced 2D-geometry, which is defined as the loci of centres of locally maximal circles inscribed inside the geometry. Medial axis together with the associated radius function form the medial axis transform (MAT) which has been used for shape recognition in image processing over a decade. In 2015, Ding et al. published a methodology for computing medial axis for any arbitrary 2D geometry and implemented it for toolpath generation of thin-wall components built using WAAM technology [104]. Figure 5.46.a. shows the medial axis transform derived for an arbitrary geometry and Figure 5.46.b. shows the offsets generated from the medial axis towards the boundary.

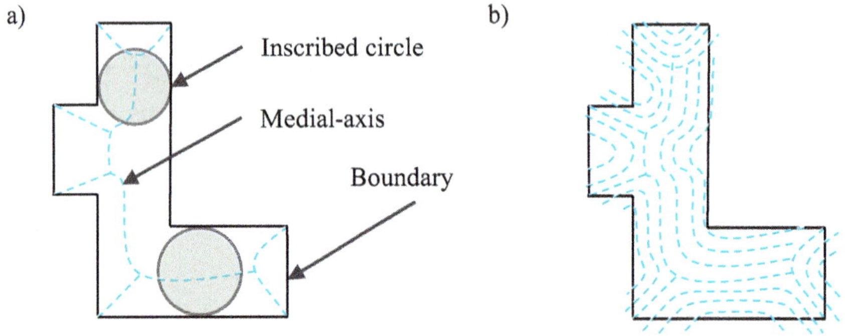

Figure 5.46 a) Medial axis transform representation of an arbitrary geometry b) Offsets generated for the medial axis from inside-out

In this research work, the medial axis transform (MAT) has been extensively used for generation of toolpaths and for topology analysis, especially for thin-wall geometries. Of the several open-source libraries that have been found for medial-axis computation, majority of them were developed for image analysis. Few python-based packages like Trimesh and Shapely provides function to compute medial axis of polygon objects. C++ based library CGAL also provides the required function for medial axis extraction of 2D and 3D tesellated geometries. Since the CAM framework developed in this work is Python-based, Trimesh library has been chosen for further implementations. Figure 5.47 shows the result of medial-axis extracted for a bulk volume object. This is, however, not suitable or practicable for deposition purposes. MAT is rather useful for path planning of thin-walled components shown in Figure 5.44.

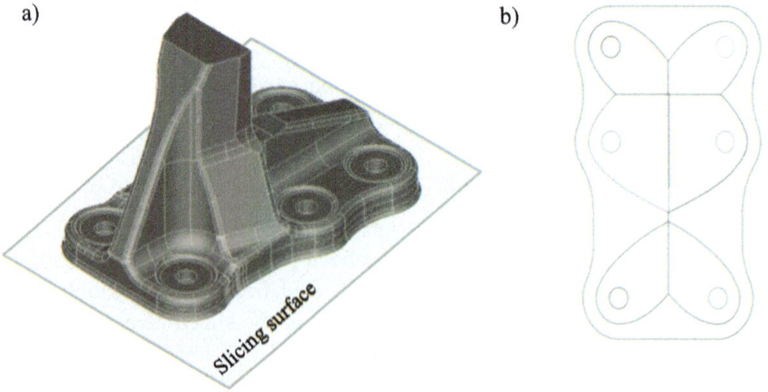

Figure 5.47 a) CAD representation of the demonstrator component shown in Figure 5.1 b) Medial axis generated for a layer sliced from the demonstrator geometry

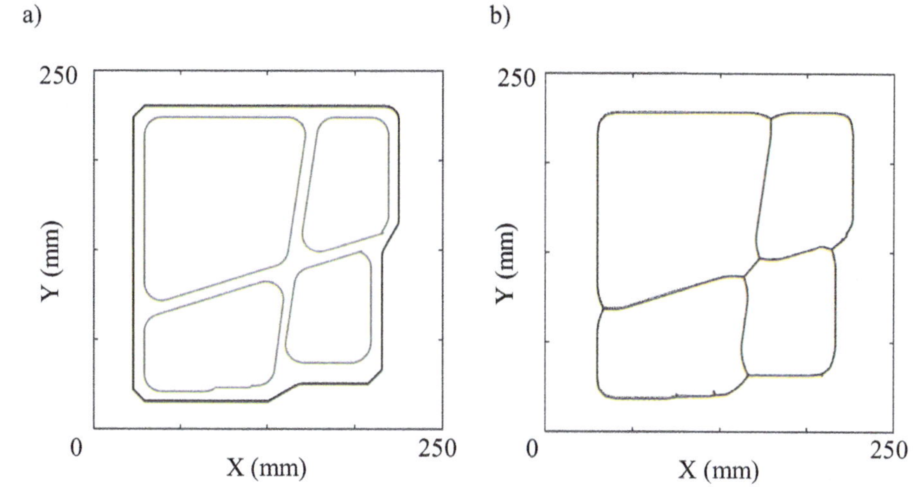

Figure 5.48 a) Representation of the sliced layer from the demonstrator geometry 1 b) Extracted medial axis of the sliced layer

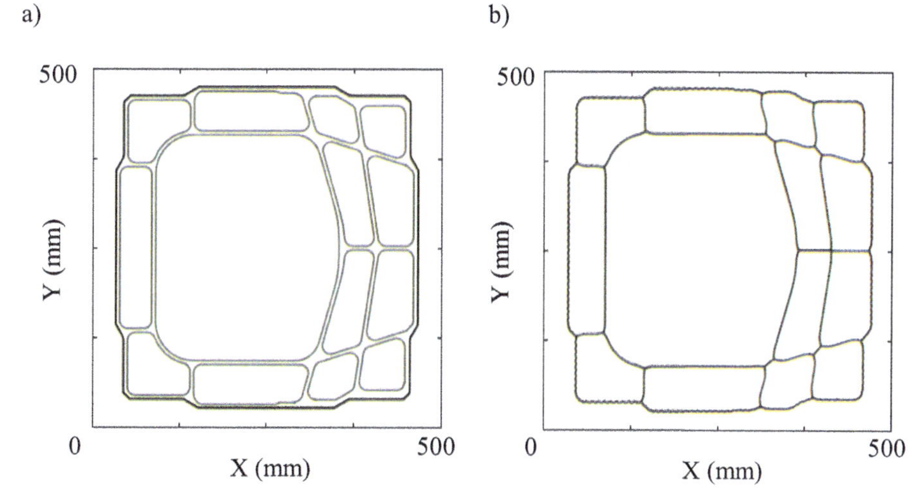

Figure 5.49 a) Representation of the sliced layer from the demonstrator geometry 2 b) Extracted medial axis of the sliced layer

Material deposition can now follow the extracted medial axis. However, if the deposited weld bead thickness is smaller than the designed wall thickness, the extracted MAT is alone not sufficient. Therefore, offsets from the medial axis (inside-out) are required in many cases. Figure 5.50 shows such offsets generated from the medial axis for the demonstrator geometry shown in Figure 5.48. Certain sections like X- and T- sections are magnified to highlight the toolpath connections and frequent change in toolpath trajectory which causes deceleration and acceleration of the robotic system at these intersections leading to inconsistant material deposition causing geometrical inaccuracies.

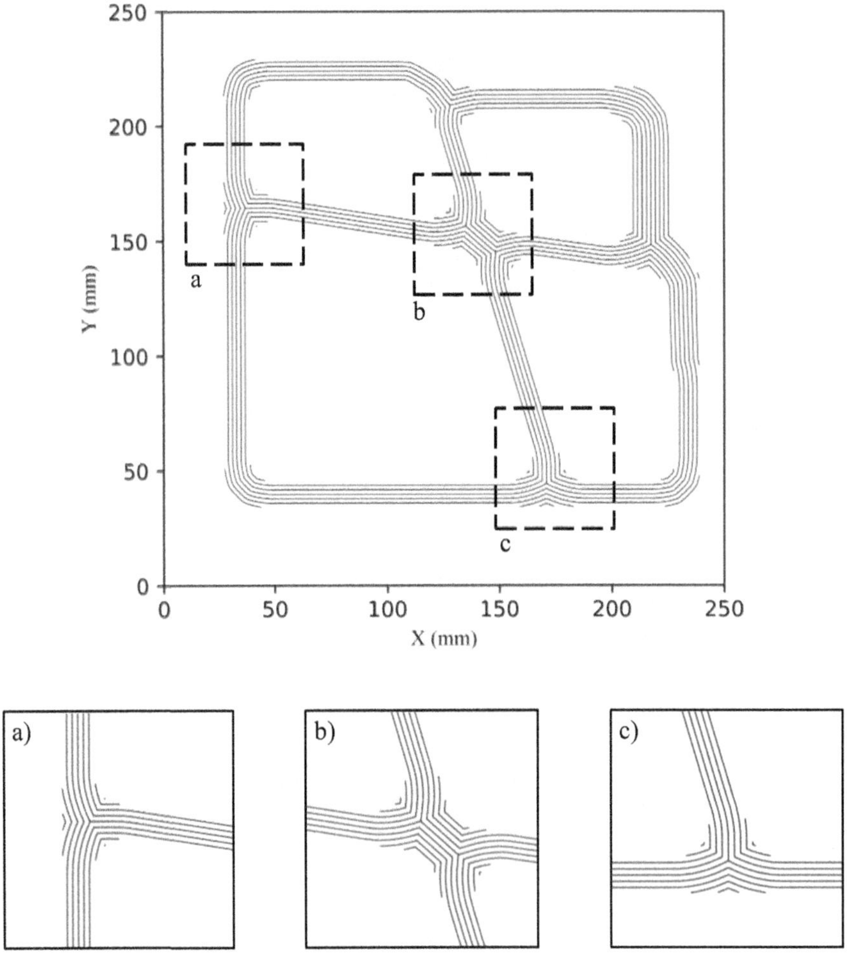

Figure 5.50 Toolpaths generated for a layer sliced from the demonstrator geometry by offsetting the extracted medial axis inside-out

The only possible way to eliminate these geometrical inaccuracies is by avoiding these critical intersections. This can be achieved by segmentation prior to extraction of medial axis. In case of this particular demonstrator, the entire component can be segmented into inner and outer features, as shown in Figure 5.51. Medial axes generated segment-wise after the segmentation process is then used for toolpath generation. Figure 5.52 shows the toolpaths generated by offsetting two medial axes separately. The figure also shows the same intersections shown previously for direct comparison. Toolpaths generated in this manner have the least number of inter-connections and also result in smooth toolpath trajectories. Such toolpaths help in achieving better geometrical accuracies.

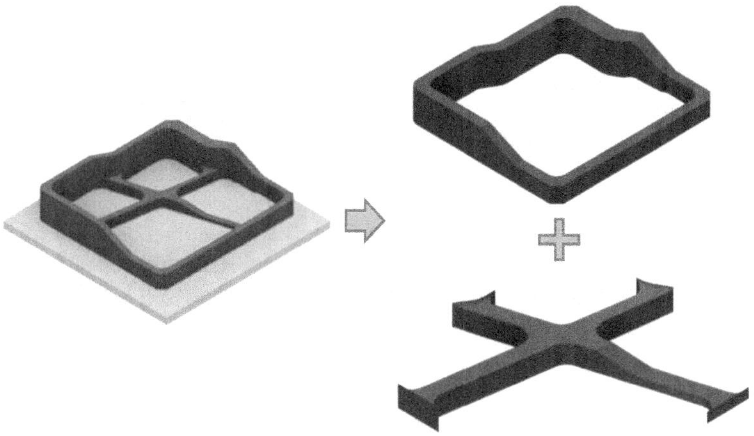

Figure 5.51 Manual segmentation of demonstrator into inner and outer geometry

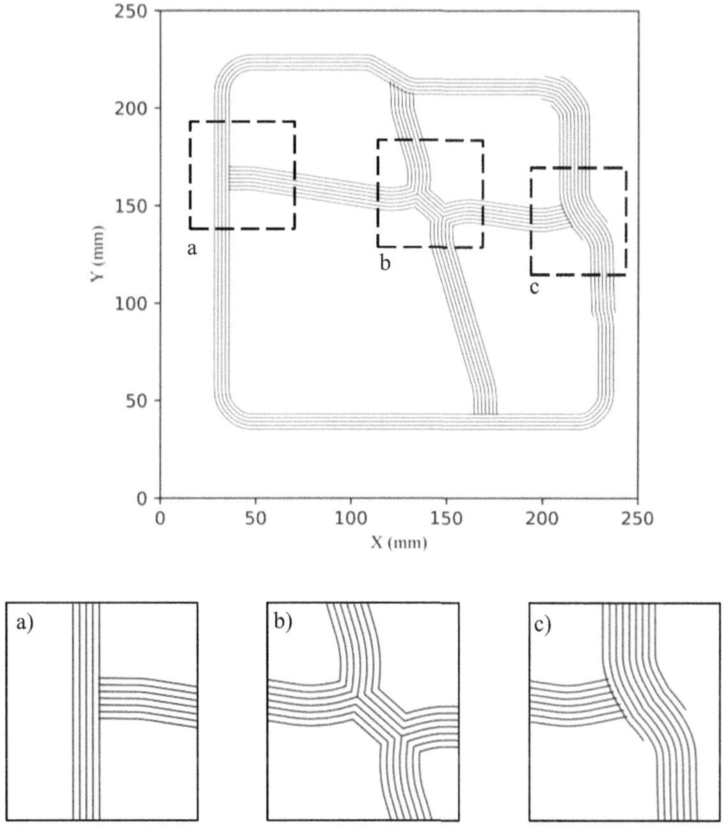

Figure 5.52 Toolpaths generated after segmentation of the demonstrator geometry into outer and inner features by offsetting the two medial axes inside-out

5.5.2. Segment-wise Toolpath Generation

Even though an alternate to the contour-based toolpath generation approach has been made possible through medial axis-based toolpath generation, for certain wire-based DED processes like WAAM process or wire-based LMD process it is necessary for process stability that the toolpaths are shorter and follow a zig-zag hatch pattern rather than the lengthier contour- or medial axis-based toolpaths [117, 118]. Hence, it is common to segment the layer into several segments followed by generation of zig-zag toolpaths to the segmented region and deposit the geometry segment-wise. Segmentation of thin-wall geometry into individual segments facilitates usage of pre-determined toolpath and build strategies for every segment type. Segmentation can be either wall-wise (detected based on geometry change) or feature-wise (detected according to intersection type). As mentioned earlier, the extraction of these walls or features from the CAD model has been predominantly a manual process carried out in the CAD environment. This section of the chapter presents the algorithms that has been developed to automate the segmentation process.

5.5.2.1. Wall-wise segmentation

Segmentation based on geometry change has been used as a technique to detect and segment walls within a sliced layer. In addition to own algorithms, methods from the 3rd party python library namely Shapely and Trimesh have been used in this function. The pseudocode for this algorithm is shown below followed by implementation results on the demonstrator geometry.

Pseudocode

Function:	Segmentation of layer based on geometry-change
Input:	STL file, slice thickness
Result:	Returns list of closed polygons

Read stl file using Trimesh's load_mesh() function

Slice the stl file with user-defined slice thickness

for i = 1 to number_of_layers

  convert slice information into polygons using Shapely function

  if polygon_area > 0

    extract inner and outer contours of the polygons

    extract the corners by determining the change of angle between adjacent contour line segments

    append both external and internal corners to a list

    determine adjacent corners by computing the distance between corners; if the distance is equal to or less than twice the average wall-thickness, then it is classified as an adjacent corner

    join adjacent corner pairs with a Shapely line object

    merge the contours and the lines and then polygonise

for j = 1 to number_of_polygons
    remove repeated polygons
    remove polygons with area = 0
  end
 end
end
Return list of closed polygons

Implementation and Results

Figure 5.53 shows the implementation result of the above-described wall-wise segmentation algorithm for the demonstrator geometry 1.

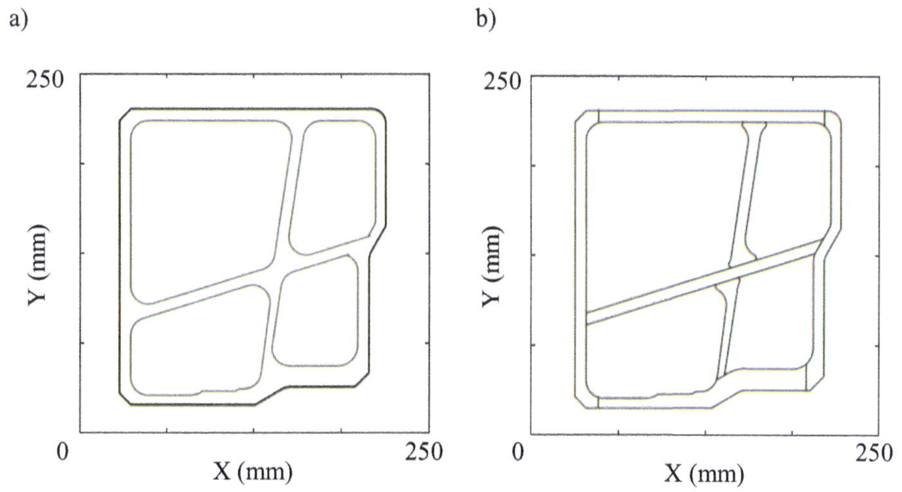

Figure 5.53 a) Representation of the sliced layer from the demonstrator geometry 1 b) Segmentation of the sliced layer based on detection of geometry change (wall-wise)

Segmentation of the sliced layer is followed by toolpath generation (or hatching) required for the deposition system. In contrast to the offset hatching pattern, one of the other in-fill hatch patterns namely zig-zag has been chosen for this demonstrator. The choice is based on the geometry and the DED process being used. Figure 5.54 shows the generated segment-wise zig-zag toolpaths generated at an angle of 45°.

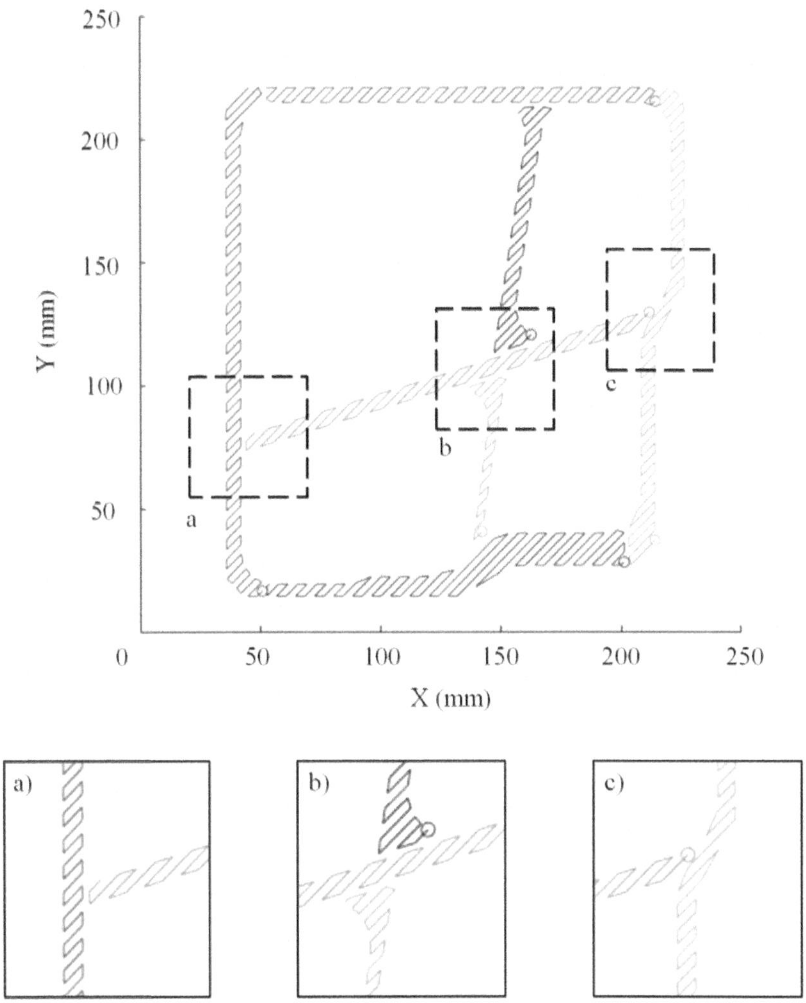

Figure 5.54 Segment-wise zig-zag toolpaths generated after segmentation of the demonstrator geometry into several walls

5.5.2.2. Feature-wise segmentation

Similar to wall-wise segmentation, an algorithm to extract pre-defined features as segments for toolpath generation has also been developed in this work. Typical features like L-shaped corners, T-, X-, and H-shaped intersections and straight-line segments can be identified based on the topology information of the medial axis extracted for the sliced layer. Toolpath strategies specific to these features can then be applied for optimum deposition results. The algorithm described below automates the manual feature-wise segmentation process carried out by engineers within the CAD environment.

Algorithm summary

The algorithm begins with the extraction of the medial axis of the sliced layer as described previously in section 5.5.1.2. This is followed by the extraction of topology information from the medial axis which is done by scanning the entire medial axis with two concentric circles of different radii for geometry change and occurrence of intersections. While scanning, the concentric circles intersect the medial axis at different locations. Line segments are created based on these intersections between the inner and outer circle. By extending these line segments to the center of the concentric circles followed by the determination of its relative orientation, information on the topology of the medial axis can be extracted (see Figure 5.56). Figure 5.55 represents the concentric circle used for scanning schematically.

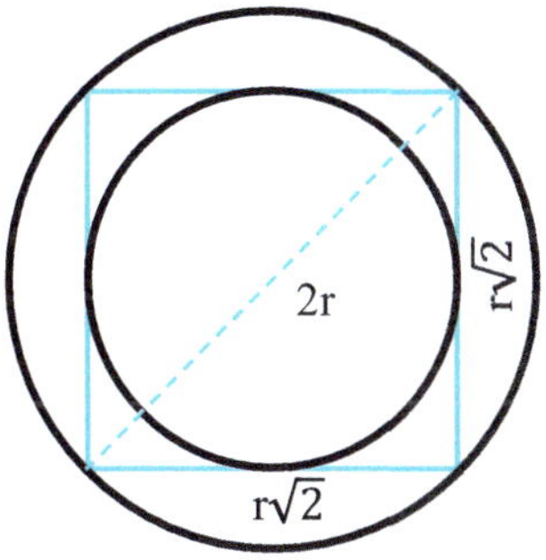

Figure 5.55 Schematic representation of the concentric circles used for scanning of the medial axis of the layer to segmented

Depending on the width of the thin-wall geometry to be scanned, the radius of the outer concentric circle is defined as such the scanning circle inscribes the feature to be detected. The outer circle has the diameter, $2r$. A square of side $r\sqrt{2}$ is inscribed inside the outer circle which is later used as a bounding box for segmenting the geometry feature, instead of using the circle itself. The inner circle is a circle that can be inscribed inside the bounding box.

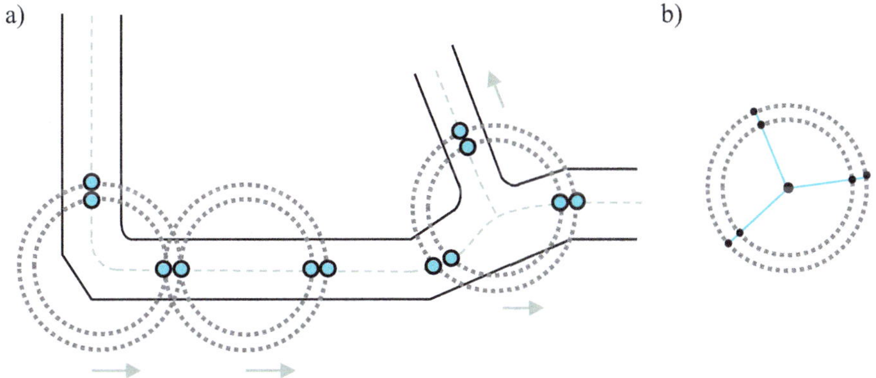

Figure 5.56 a) Schematic representation of the two concentric circles scanning the medial axis b) Line segment creation and extension to the center of the circle for orientation determination

Algorithm flowchart

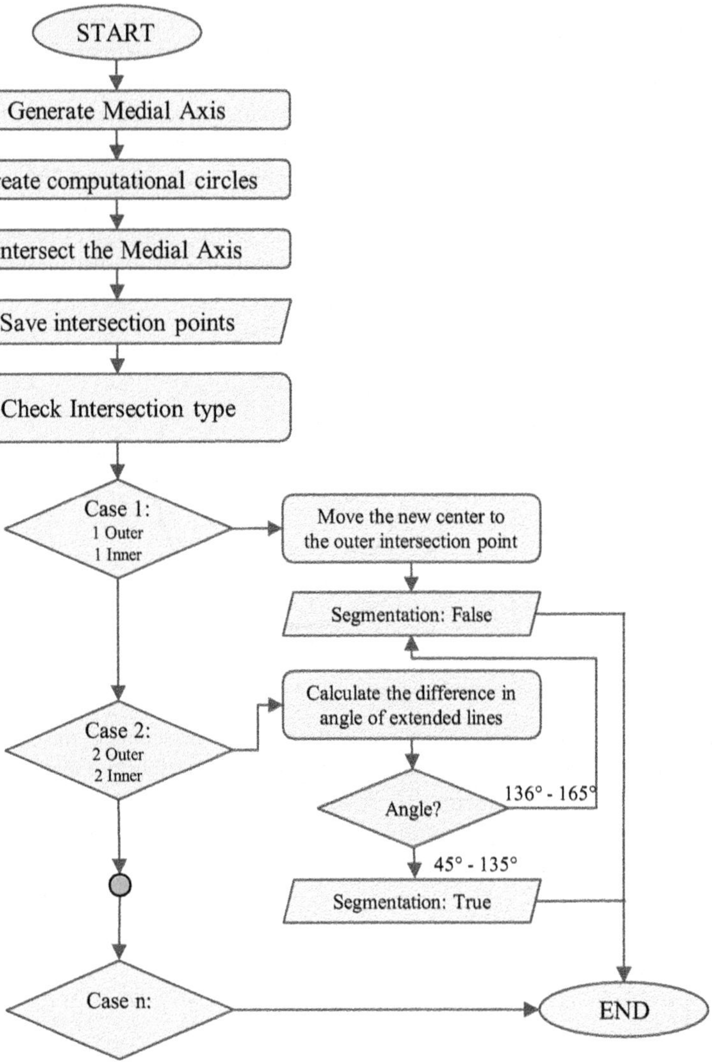

Figure 5.57 Flowchart showing the working principle of feature detection and segmentation algorithm

Figure 5.57 shows the working of the scanning, feature detection and segmentation algorithm as a flowchart. Several pre-determined intersection conditions are programmed as if else statements. Feature detection is an iterative scanning process starting at (X_{min}, Y_{min}) of the geometry. After successful detection of the first feature by satisfying one of the several angular conditions, the location information of the concentric circles as well as the bounding box inscribed within the outer circle is stored. All possible scan directions are stored as well after the first iteration derived from the location of the intersection points of the concentric circles with the medial axis. In the following iteration, the scanning continues until another feature is detected simultaneously along the directions list from the previous iteration. This iteration continues until all possible scan directions have been searched and all possible features have been extracted. Scan duplicities are avoided by removing the detected feature for the next iteration. A schematic representation can be seen in Figure 5.58 for the demonstrator geometry shown in Figure 5.44.a. For this particular demonstrator geometry, all the features have been detected after 5 iterations.

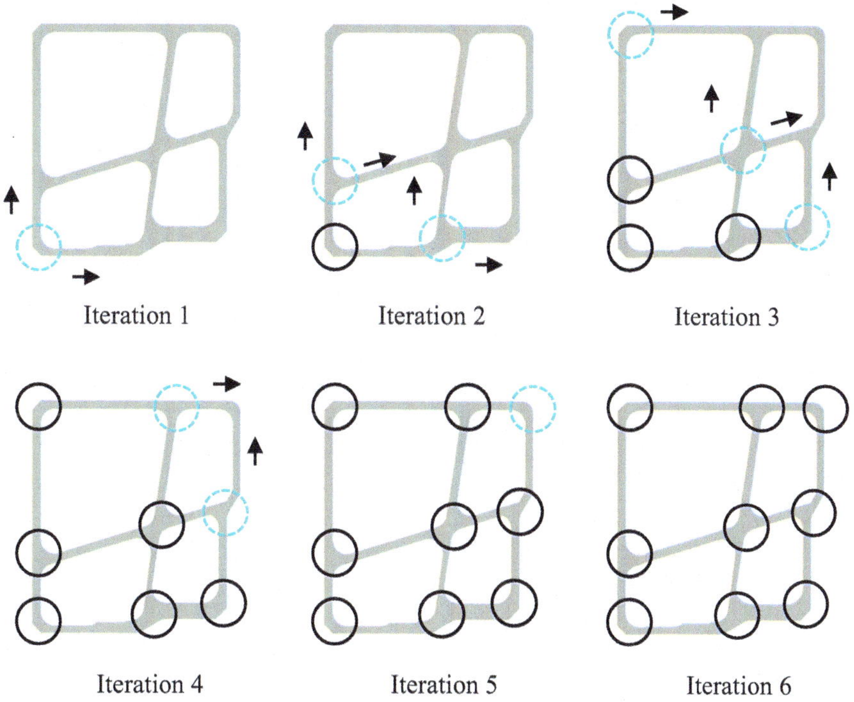

Figure 5.58 Schematic representation of the iterative scan and feature detection process developed for feature-wise toolpath generation approach

Implementation and results

Implementation of the feature recognition and segmentation algorithm for demonstrator geometries can be seen in Figure 5.59. and Figure 5.60. Figure 5.61 depicts the toolpath generated for the different features segmented in the previous step. Pre-tested feature-wise toolpath strategy can be implemented for optimum deposition results, however, for demonstration purpose a single 45° zig-zag infill pattern has been implemented for all segmented features.

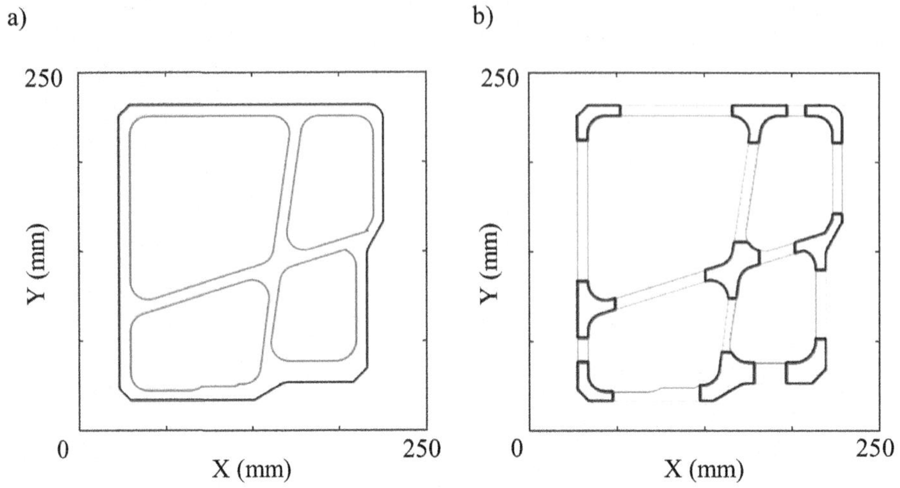

Figure 5.59 a) Representation of the sliced layer from the demonstrator geometry 1 b) Segmentation of the sliced layer based on feature detection

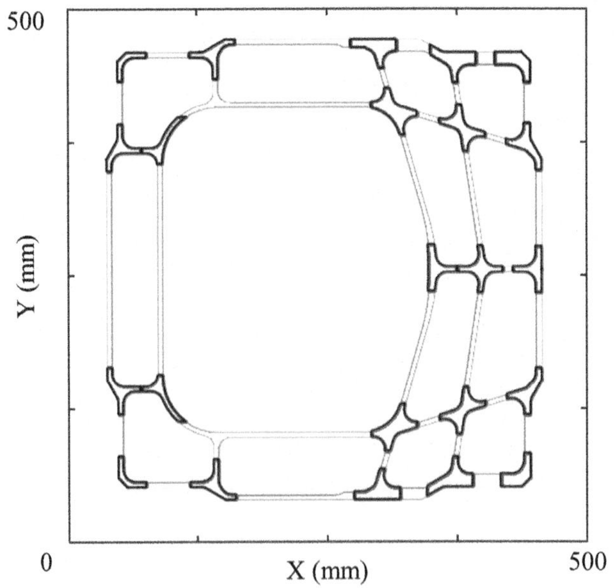

Figure 5.60 Segmentation of the sliced layer of demonstrator geometry 2 based on automatically detected features

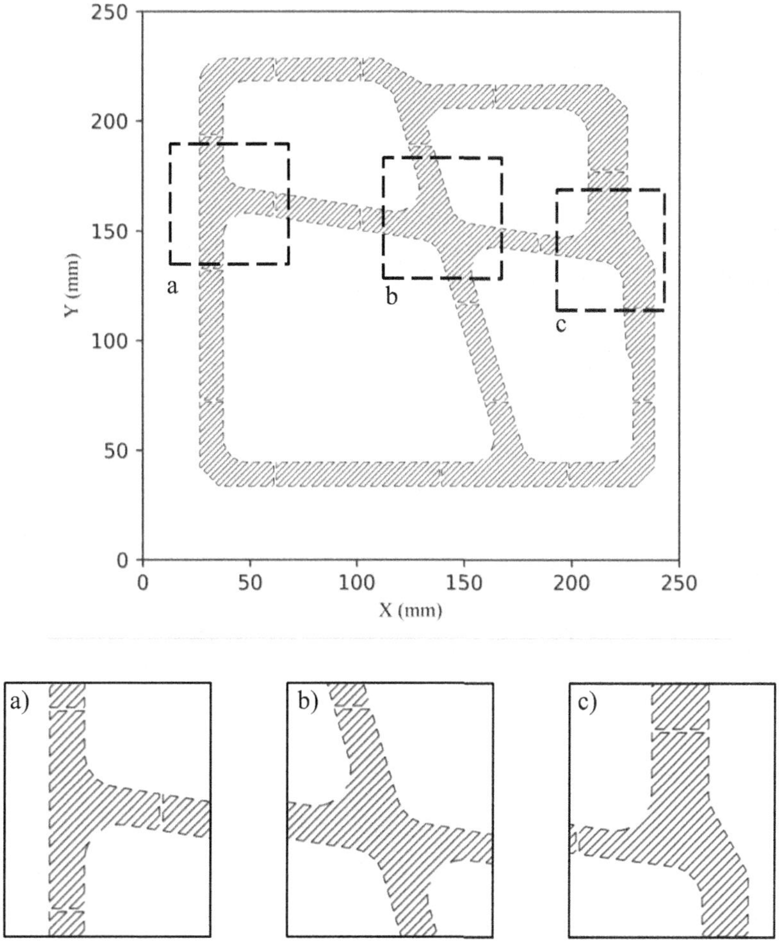

Figure 5.61 Feature-wise zig-zag toolpaths generated after segmentation of the demonstrator geometry into several features

5.6. Process Planning Platform

The algorithms developed in pyhton for every component cluster can be executed individually in a python environment. The required user-defined variables are either provided in the code or received through terminal. There exists no graphical user interface (GUI). The developed algorithms can be adapted and implemented in existing commercial software packages or even used to develop a new open-source platform for manufacturing process planning. Such a development is beyond the scope of this research work.

However, a prototype of such a process planning platform using the same python framework has been proposed in this section. The framework, with the associated libraries and packages, that was used to develop the prototype is shown in Figure 4.2, without much information on the software architecture, class structure and the codes used. Figure 5.62

and Figure 5.63 shows snapshots of the prototype where the algorithm for STEP-file format based planar slicing has been implemented.

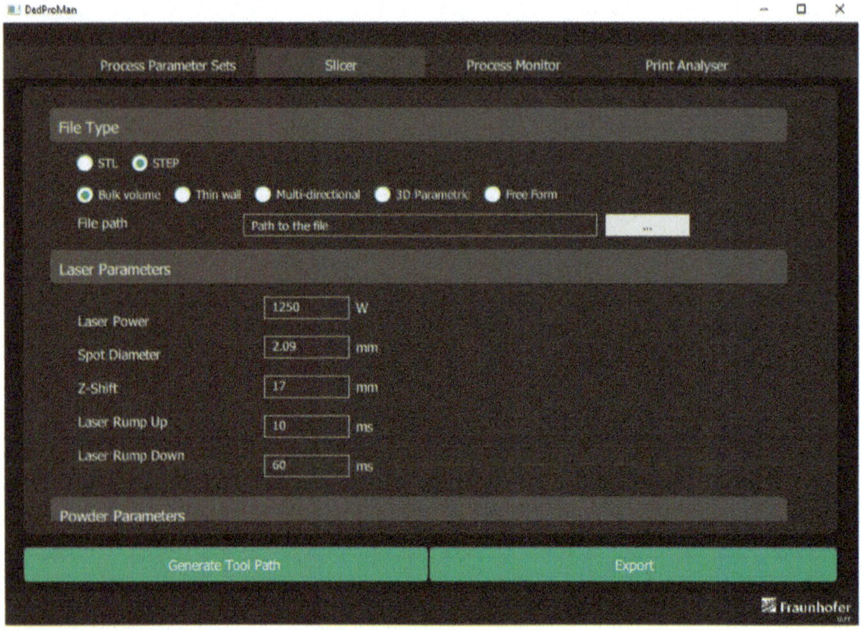

Figure 5.62 Graphical user interface of the process planning platform

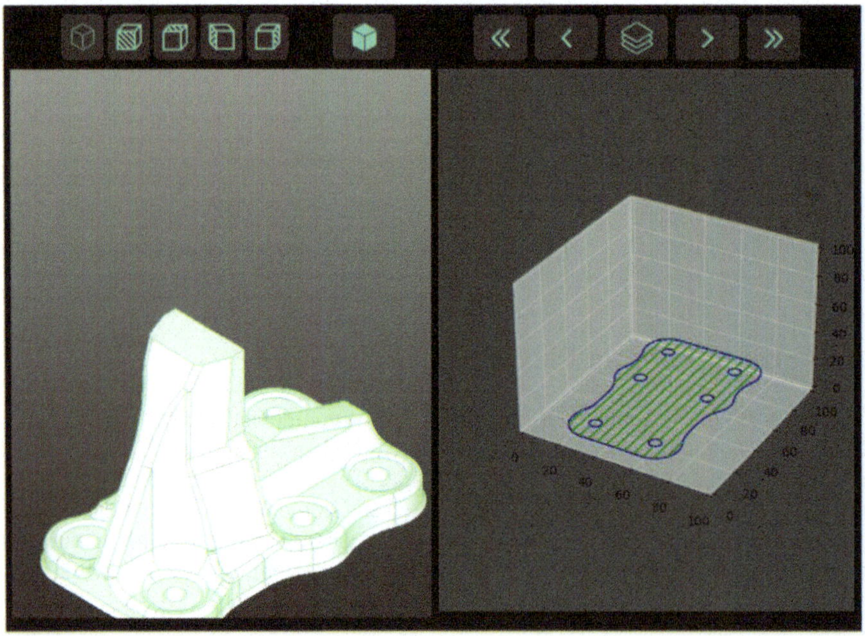

Figure 5.63 3D and 2D viewer of the process planning platform for STEP data and toolpath visualization respectively

6. Influence of Process Parameters and Toolpath Strategy on Build Quality

The resulting geometrical, metallurgical and mechanical characteristics of the DED part depends upon the cooling rate and the resulting microstructure of the individual layers and further more upon individual weld tracks. The cooling rate is essentially governed by several intrinsic and extrinsic process parameters and the process mechanics itself. Additionally, component geometry, toolpath and build strategy, substrate geometry, clamping forces and many such factors also influence the resulting metallurgical characteristics[149, 150, 131]. In addition to these controllable process parameters, the interaction between different system components and parameters also influences the occurring physical phenomena and melt pool dynamics. A particular process parameter set (laser power, deposition velocity and powder mass flowrate) might yield different results with two different toolpath strategies or different substrate dimensions due to the different heat transfer mechanisms [151, 152].

Even though numerous thermo-mechanical simulation models have been developed to predict the melt pool dynamics, the resulting temperature field and the deformation, they prove to be useful only for simpler geometries. Several experiments are necessary for validating the model. Moreover, these models are developed under the assumption that a constant working distance between the nozzle and substrate (or pre-built layers) is maintained over the build direction and therefore a constant weld bead geometry is deposited. This, however, is not the reality. In addition to the primary process parameters, component geometry, toolpath strategy, substrate dimensions and other secondary parameters decide the resulting weld bead dimensions, which are not considered in the simulation models. Therefore, in industrial setup, often an experimental approach is the chosen methodology, which is essentially a trial-and-error approach. This works also uses an experimental approach to identify the primary and secondary process parameters and also to identify the correlation between them. However, instead of a trial-and-error approach, a systematic methodology has been proposed thereby making the process lean. Section 6.1 describes the system setup and the particular DED technology chosen for this purpose. Section 6.2 presents the methodology developed for systematic identification of primary and secondary process parameters with test geometries. In section 6.3, the developed methodology has been evaluated with industrial use-cases.

6.1. Process Setup and Material

The DED process chosen for implementation and verification of the developed toolpath strategies and for the systematic identification and evaluation of process parameters was powder-based laser metal deposition process. For this purpose, the TRUMPF TruLaser Robot Series 5020 equipped with a high-precision 6-axis industrial robot KR 60 HA from KUKA AG was used (see Figure 6.1). For deposition of freeform geometries, this manufacturing robot cell is equipped with a 3-nozzle deposition head with a powder feeder unit that delivers metal powder to the nozzle head. A maximum of four powder containers could be placed in the hopper unit and each powder hopper is equipped with a rotating

V. J. Prakash, *Automated Process Planning for Additive Manufacturing with Directed Energy Deposition*, Light Engineering für die Praxis,
https://doi.org/10.1007/978-3-031-87666-0_6

disk with a machined slot to transport powder particles from the hopper through rotation. These powder particles are then transported to the nozzle by a carrier gas (usually Helium) through a pneumatic tube. The quantity of the metal powder to be delivered is then controlled by the rotational speed of the circular disk and the slot area in the circular disk. TRUMPF TruDisk 6001 is the laser energy source used for this system setup, which is a diode-pumped disk laser system with a maximum output power of 6 kW. Yb:YAG is the active laser medium that generates laser at a wavelength of 1030 nm. Technical specifications of the described system setup are tabulated in Table 6.2.

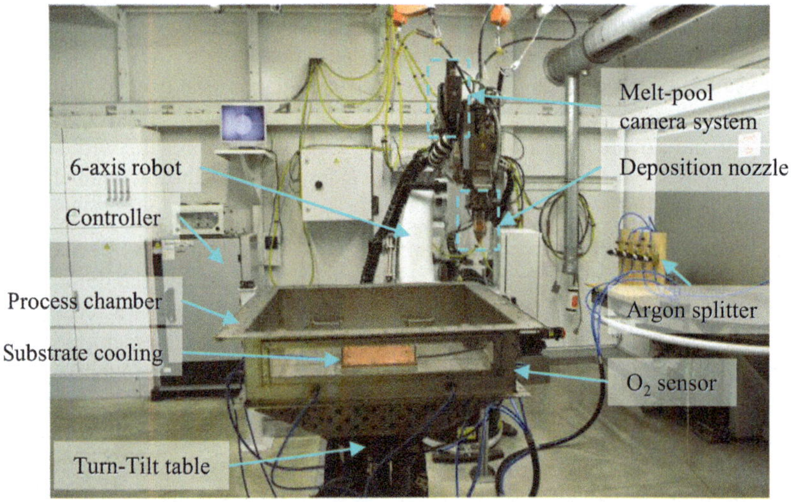

Figure 6.1 Laser metal deposition system setup with in-house fabricated inert gas chamber for deposition of reactive materials

Material used in this research work was one of the most widely used titanium alloy in the aerospace industry, Ti-6Al-4V. This α+β Ti-alloy provides a combination of good specific material properties along with an excellent corrosion resistance. Higher material costs and higher machining costs of Ti-alloys makes this material an excellent choice for development with additive manufacturing. Furthermore, higher buy-to-fly ratio of aerospace structural components leads to titanium scraps that are not recycled till date. Material specifications of the Ti-6Al-4V powder used for demonstration fabrication is shown in Table 6.1.

Table 6.1 Ti-6Al-4V powder specifications

Material parameter	**Unit**	**Value/ Range**
Alloy	-	Ti-6Al-4V
Grade	-	ASTM Gd 5
Atomization	-	Gas (Ar)
Powder size	μm	50 -100
Powder shape	-	spherical
Flowability	s/50 g	< 28 s/50 g

Table 6.2 Technical specifications of the LMD system setup

System parameter	Unit	Value/ Range
Laser generator		
Source type	-	Yb:YAG disk laser
Power	W	500 - 6000
Wavelength	nm	1030
Fiber diameter	µm	200
Beam quality	mm. mrad	4
Powder delivery system - Deposition Line		
Powder containers	-	2
Powder mass flow rate	g/min	≤ 16.5
Carrier gas	l/min	≤ 20
Shield gas	l/min	≤ 20
Fiber diameter	µm	100
Focal length	mm	150
Focal diameter	mm	0.4 - 6
Positioning system - KUKA robot KR 60 HA		
Number of axis	-	6 + 2
Reachability (max)	mm	2429
Weight	kg	655
Repeatability	mm	± 0.05
Payload (max)	kg	30
Controller	-	KR C2

6.2. Systematic Identification of Process Parameters

In this section, the systematic approach developed to identify the primary and secondary process parameters and their correlation through experimental findings has been presented. Figure 6.2 shows the schematic overview of the methodology developed.

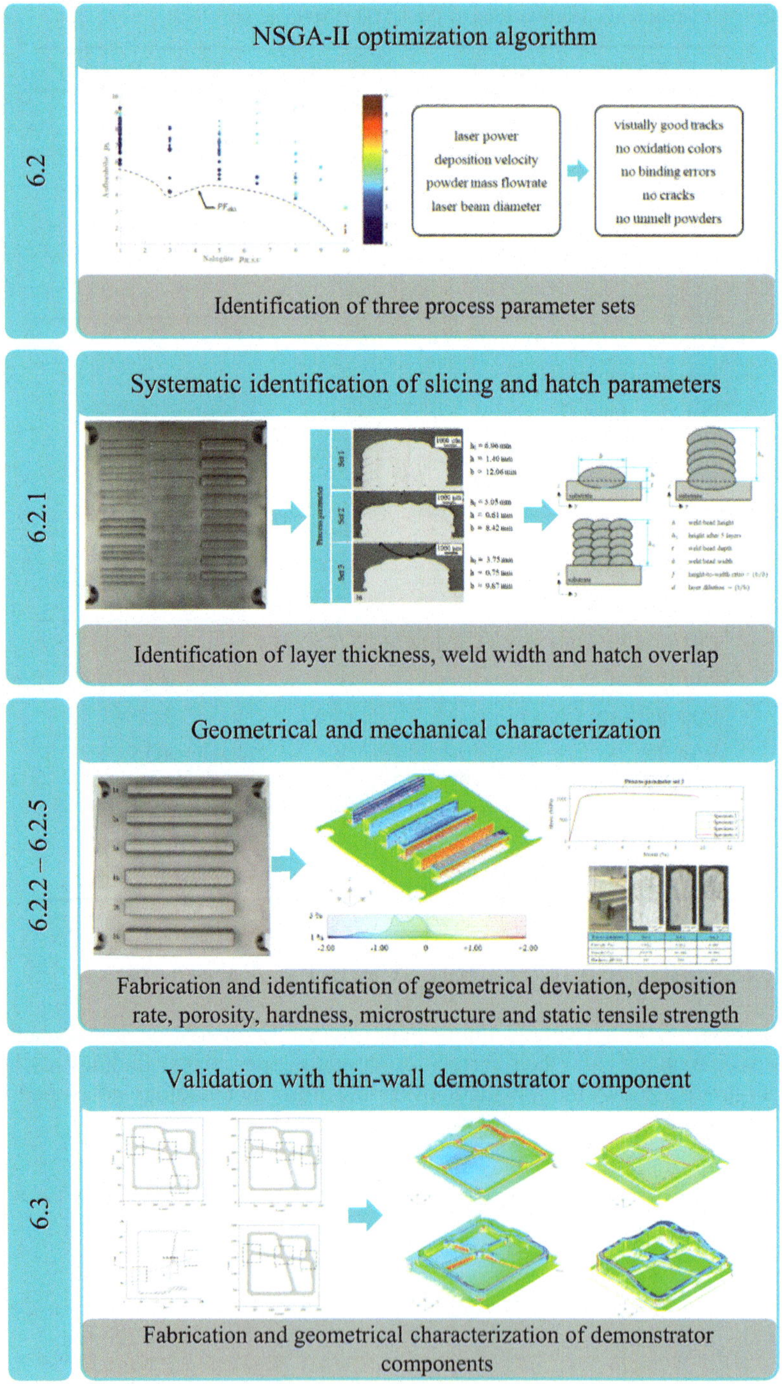

Figure 6.2 Schematic representation of the methodology developed for systematic identification and evaluation of process parameters and toolpath strategies

6.2.1. Experimental Identification of Slicing and Hatch Parameters

Process optimization begins with the selection or optimization of primary process parameters including laser power, deposition velocity, powder mass flow-rate and laser beam diameter. These factors are optimized through either classical full-factorial DoE (design of experiments) which is time and resource consuming or through multi-objective optimization algorithms like NSGA-II (Non-dominated Sorting Genetic Algorithm) and RSM (Response Surface Methodology). Usage of pareto optimization techniques reduces the total number of experiments to be executed, thereby reducing the resources required. Extensive work on process parameter optimization through NSGA-II methodology has already been carried out with the system technology mentioned in section 6.1. using Ti-6Al-4V powder [153, 154]. In the mentioned work, the aforementioned primary process parameters were optimized for multiple objectives like aspect ratio of the weld bead geometry, bonding with base plate (visual inspection and contact angle) and crack-free and spatter-free weld beads. From the work, three process parameter sets have been chosen for further analysis which are tabulated in Table 6.3.

Table 6.3 Three chosen process parameter sets with their primary process parameters and some secondary parameters

Process parameter	Unit	Process parameter set		
		1	2	3
Laser power, P_l	W	1450	700	1500
Deposition speed, U	m/s	0.01	0.01	0.0267
Powder mass flow rate, $\dot{m}$	g/min	13.10	6.00	16.10
Spot diameter, d_{spot}	mm	2.03	1.50	2.03
Intensity, E	W/mm^2	424.27	396.11	437.28
Energy per unit length, $\dot{E}$	kJ/m	145.00	70.00	56.25

The chosen process parameter sets are then combined with two different toolpath strategies, namely parallel and zig-zag toolpaths, as represented schematically in Figure 6.3. These two different toolpath strategies replicate two extreme cooling rate phenomena. For the two spatially adjacent points in adjacent toolpaths (marked as red points in the figure), the temporal space between them during deposition is different. The temporal space in parallel toolpath is much higher than that of the zig-zag toolpath, resulting in a different melt pool dynamics, cooling and solidification rate, and the resulting geometrical dimensions in comparison to the zig-zag toolpath.

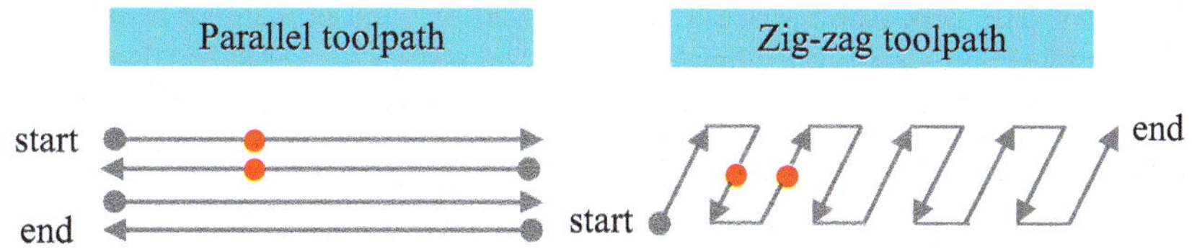

Figure 6.3 Schematic representation of parallel and zig-zag toolpath strategies

The experimental plans to deposit single weld lines, single walls, surfaces and blocks for parallel and zig-zag toolpaths are shown schematically in Figure 6.4 and Figure 6.6. The process parameter sets are differentiated by color. Three test specimens were deposited for single weld lines and single wall geometries, whereas only one specimen was deposited for surface and block geometries. The surface and block geometries include an additional hatching parameter called ‘hatch overlap’. 30%, 35% and 40% hatch overlap values were chosen in the experimental plan. The specimens were numbered from 1 to 54, which are referenced in the cross-section micrographs as well.

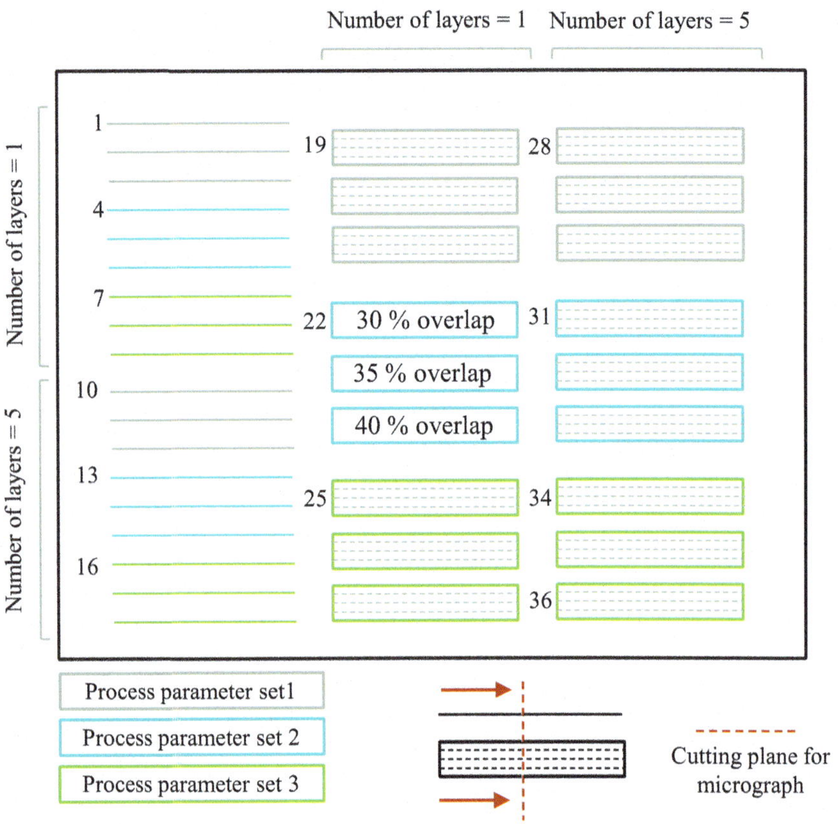

Figure 6.4 Experimental plan for single weld bead, single wall, surface and block geometries for parallel toolpath strategy

The test geometries were deposited with the system setup and material as mentioned in section 6.1. The deposition process was followed by sand blasting to remove adhered powder particles and stress-relief heat treatment according to AMS2801 [155] under air atmosphere. The build job was once again sand blasted to remove the brittle oxide layer that forms during heat treatment under atmospheric air. The surface however would still possess a thin layer of oxide in μm depth that cannot be removed through sand blasting. Figure 6.5 and Figure 6.6.b shows the photograph of the deposited geometries with the base plate after the post-processing steps.

After the deposition process, each and every test geometry is prepared for cross-section micrographs. Using wire-cut EDM process, test geometries along with 1 mm of base plate material is removed and cut laterally to smaller sizes for further micrograph preparation. The test geometries are embedded and polished with no evidence of scratch marks. The embedded geometries are then observed under microscope for geometrical characterization, porosity measurement and defect analysis. Figure 6.7 shows a schematic and abstract representation of the test geometries and the geometrical characteristics that were measured from the micrographs.

Figure 6.5 Laser metal deposited test geometries based on experimental plan shown in Figure 6.4 on a 35 mm Ti-6Al-4V base plate, followed by heat treatment (AMS 2801 under air atmosphere) and sandblasting

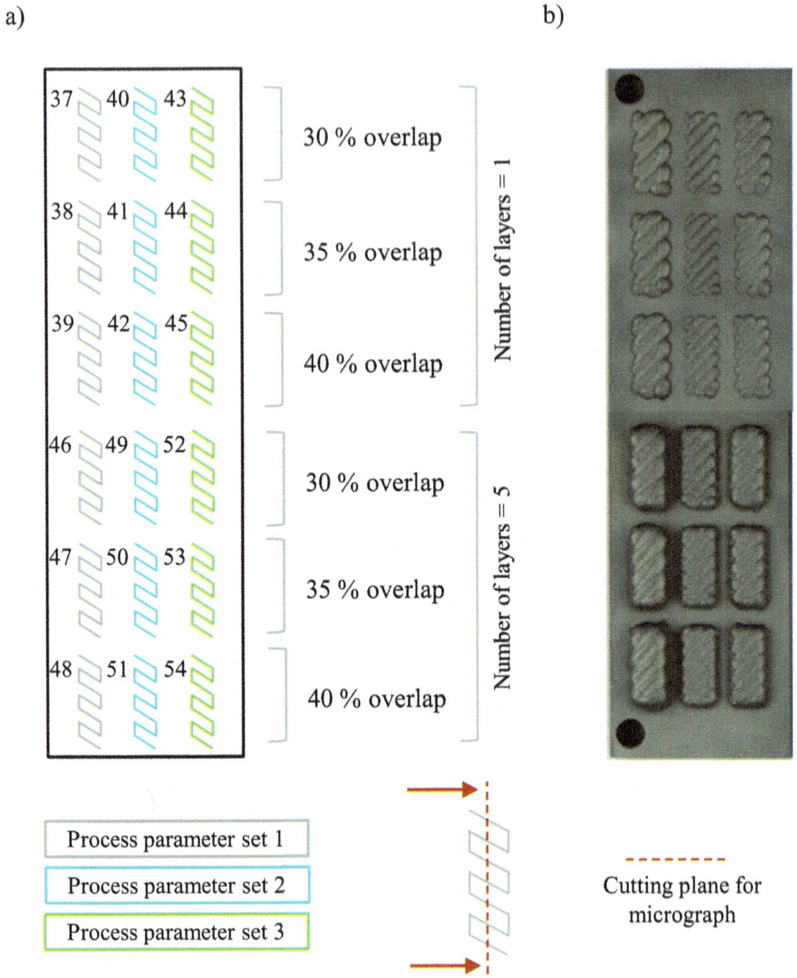

Figure 6.6 a) Experimental plan for single weld bead, single wall, surface and block geometries for zig-zag tool-path strategy b) Laser metal deposited test geometries based on experimental plan followed by heat treatment (AMS 2801 under air atmosphere) and sandblasting

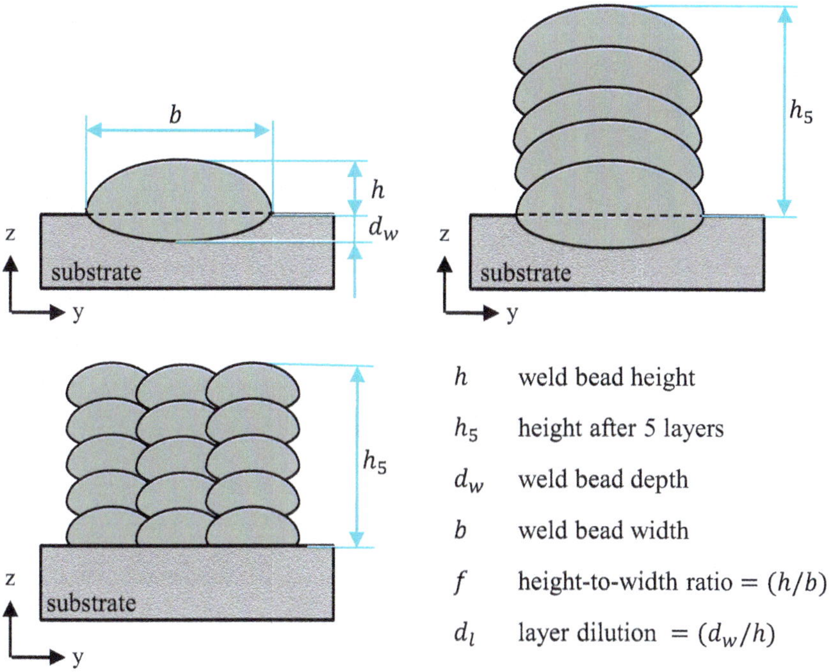

Figure 6.7 Abstract representation of the test geometries with the nomenclature of the measured geometrical parameters

Cross-section examination of the micrographs begins with single bead geometries for determination of bead height h, width b, and the resulting aspect ratio f. Figure 6.8 also shows the etched specimens from which the weld bead depth, d_w is measured and the resulting layer dilution, d_l is deduced. Weld bead height and width measured from single bead test geometries cannot be used as the slice thickness parameter due to its 3-dimensional heat transfer possibilities with the base plate leading to higher cooling rates in comparison to the to be deposited top layers. When deposited over a base plate which is maintained at room temperature, the first layer has a very large surface area for heat dissipation and therefore a steep cooling rate leading to higher weld bead height and lower width. Therefore, these values are not chosen as slicing parameters. However, the deduced aspect ratio and dilution are important quality characteristics of a parameter set. For deduction of layer thickness and weld bead width, micrographs of single wall test geometries were used. The measured total height of the wall geometry and the number of layers are used to derive the layer thickness or slice thickness for the surface and volume geometry (see Figure 6.9).

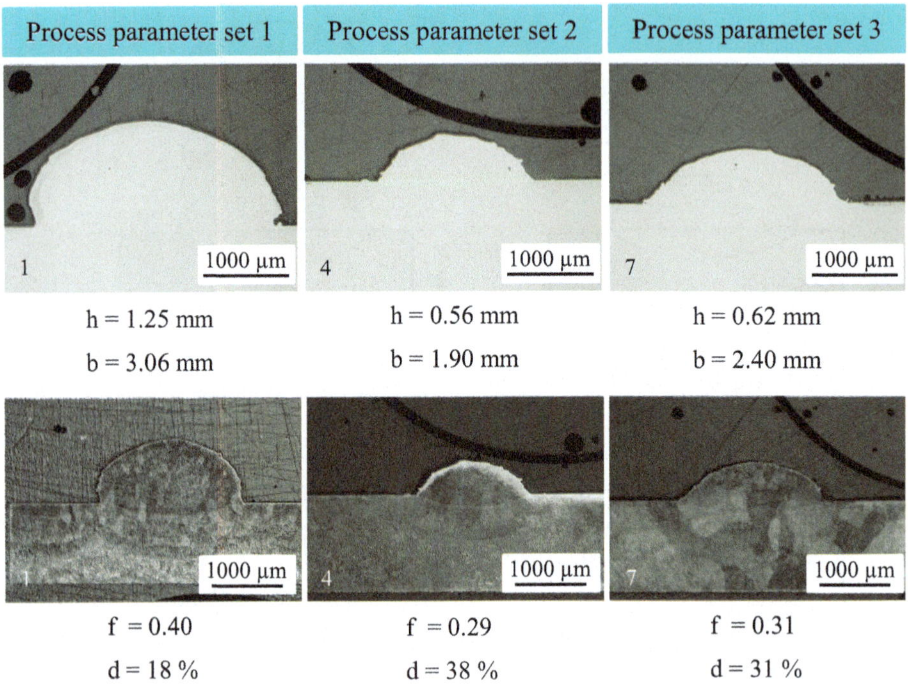

Figure 6.8 Cross-sectional micrographs of single-track weld specimens for three different process parameter sets: unetched (above) and etched (below)

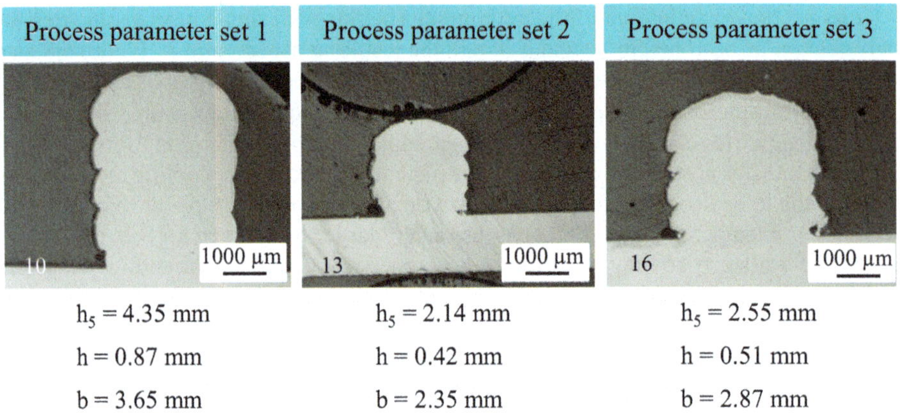

Figure 6.9 Cross-sectional micrographs of single-wall weld specimens for three different process parameter sets along with the measured geometry parameters

The derived layer thickness and weld bead width are used for toolpath preparation and deposition of surface and volume geometries with three process parameter sets and two toolpath strategies. For surface and volume geometries, an additional hatch overlap parameter was introduced with values 30%, 35% and 40%. Figure 6.10 and Figure 6.11 show the micrographs of volume geometries (with 40% hatch overlap) from parallel and zig-zag toolpath strategies respectively. Similar to the single wall geometries, the actual single layer thickness is deduced from the total build height and the number of layers deposited. Micrographs are then analyzed for porosity levels, intra and inter-layer binding defects, lack of fusion, trapped gas pores and hot cracks. Binding defects are visible in the test geometries 30 and 48, both deposited with process parameter set 1 and with 40% hatch overlap. The location and pattern of the defect shows that the underlying reason is the contact angle (acute) of the single weld bead. Cross-section micrographs of other process parameter sets are tabulated in Appendix A.2.

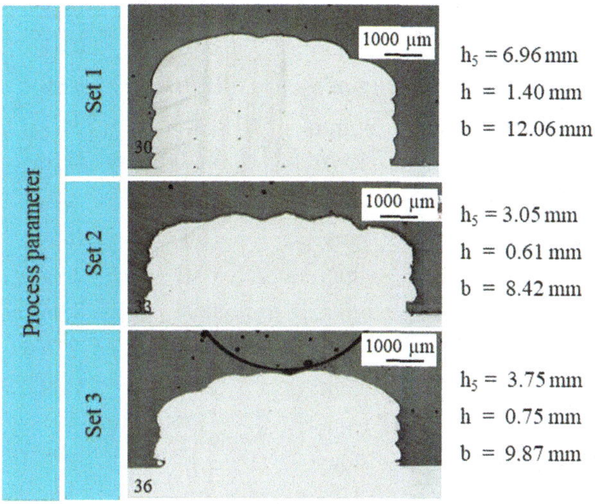

Figure 6.10 Cross-sectional micrographs of block geometry with parallel toolpaths for three different process parameter sets along with the measured geometry parameters

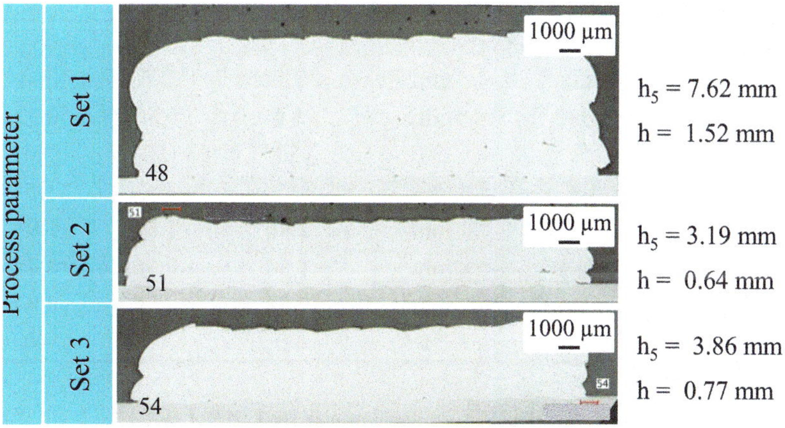

Figure 6.11 Cross-sectional micrographs of block geometry with zig-zag toolpath for three different process parameter sets along with the measured geometry parameters

After the examination of cross-section micrographs, all the required primary and secondary process parameters along with slicing and hatch parameters were finalized and are tabulated in Table 6.4.

Table 6.4 Process, slicing and hatch parameters for three process parameter sets

Parameter	Unit	Process parameter set		
		1	2	3
Process parameters				
Laser power	W	1450	700	1500
Deposition speed	m/s	0.01	0.01	0.0267
Powder mass flow rate	g/min	13.10	6.00	16.10
Intensity	W/mm^2	424.27	396.11	437.28
Energy per unit length	kJ/m	145.00	70.00	56.25
Laser ramp up	ms	10	10	10
Laser ramp down	ms	60	60	60
Weld velocity	m/s	0.01	0.01	0.0267
Non-weld velocity	m/s	0.04	0.04	0.04
Disk rotation	rpm	5.7	2.7	7
Beam diameter	mm	2.09	1.50	2.09
Z-shift	mm	17	12	17
Carrier gas flowrate (He)	lpm	4	2.7	5
Shielding gas flowrate (Ar)	lpm	20	20	20
Powder pre-flow rate	s	22	22	22
Powder post-flow rate	s	5	5	5
Dwell time	s	5	5	5
Slicing and hatch parameters				
Single weld bead width	mm	3.7	2.6	3.0
Single weld bead height	mm	1.0	0.50	0.52
Path overhang	%	50	50	50
Hatch overlap	%	35	35	35
Step-over distance*	mm	2.40	1.69	1.95
Slice thickness (Parallel)*	mm	1.30	0.56	0.70
Slice thickness (Zig-zag)*	mm	1.50	0.70	0.84
Intersection overlap	%	90	90	90

*for 35% hatch overlap

6.2.2. Fabrication of Test Geometries

With the final process parameters including slicing and hatch parameters, block geometry of dimension 150 x 18 x 20 mm^3 (L x B x H) was fabricated with three different process parameter sets and two toolpath strategies to perform geometrical and mechanical characterization. Experimental plan and the toolpath strategy used for deposition are represented schematically in Figure 6.12 and Figure 6.13 respectively. The test geometries were deposited on a 250 x 250 x 35 mm^3 Ti-6Al-4V base plate.

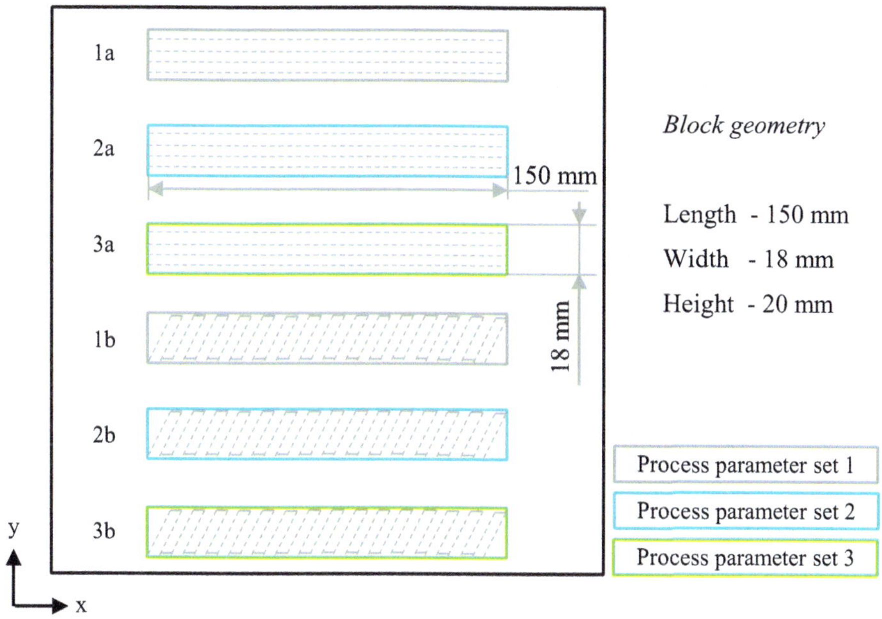

Figure 6.12 Experimental plan for deposition of block geometries with three process parameter sets using parallel and zig-zag toolpath strategies

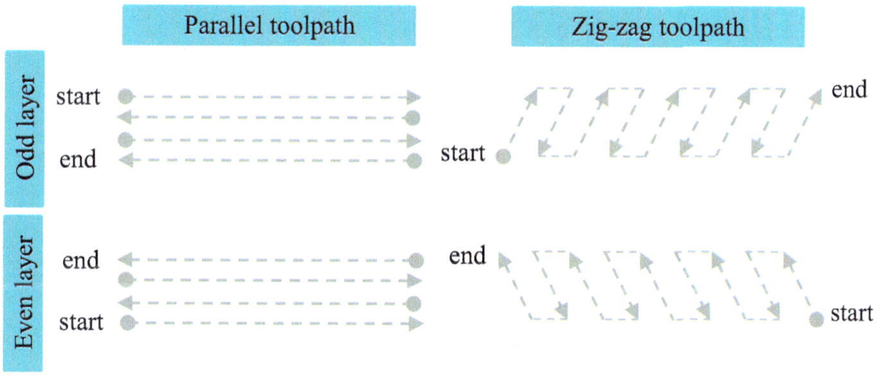

Figure 6.13 Toolpath strategy used fabrication of block geometries

Figure 6.14 shows the as-built photograph of the test geometries. It can be seen that test geometries (3a and 3b) fabricated with process parameter set 3, having the highest energy intensity, show oxidation coloration. In specific, the effect is maximum with the combination of zig-zag toolpath strategy and process parameter set 3. The oxidation decreases with the decreasing energy intensity of the process parameter. Process parameter set 2 with the least energy intensity shows no evidence of oxidation with both the toolpath strategies.

Figure 6.14 As-built photograph of laser metal deposited Ti-64 block geometries based on the experimental plan shown in Figure 6.12

Figure 6.15 shows the same test geometries after stress-relief heat treatment and sand blasting. The figure clearly shows the different toolpath patterns used for deposition.

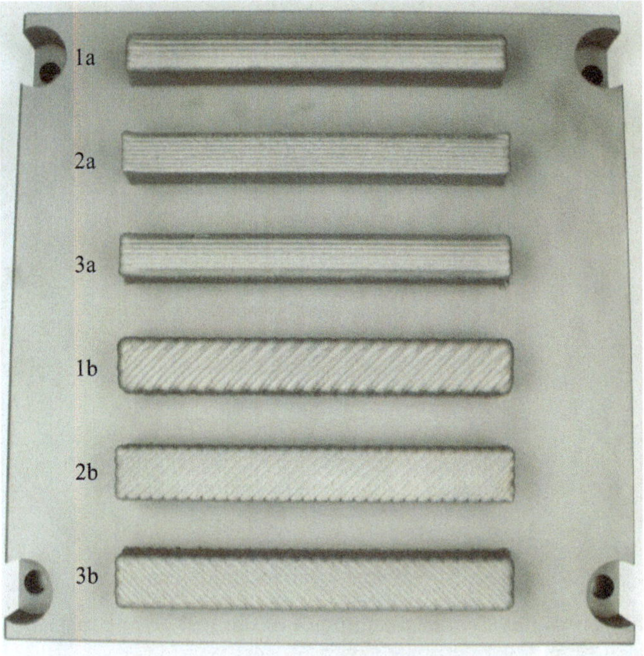

Figure 6.15 Photograph of test specimens after heat treatment (AMS 2801 under air atmosphere) and sand blasting

6.2.3. Geometrical Characterization

After heat treatment and sand blasting of test geometries, the build job was subjected to geometrical characterization.

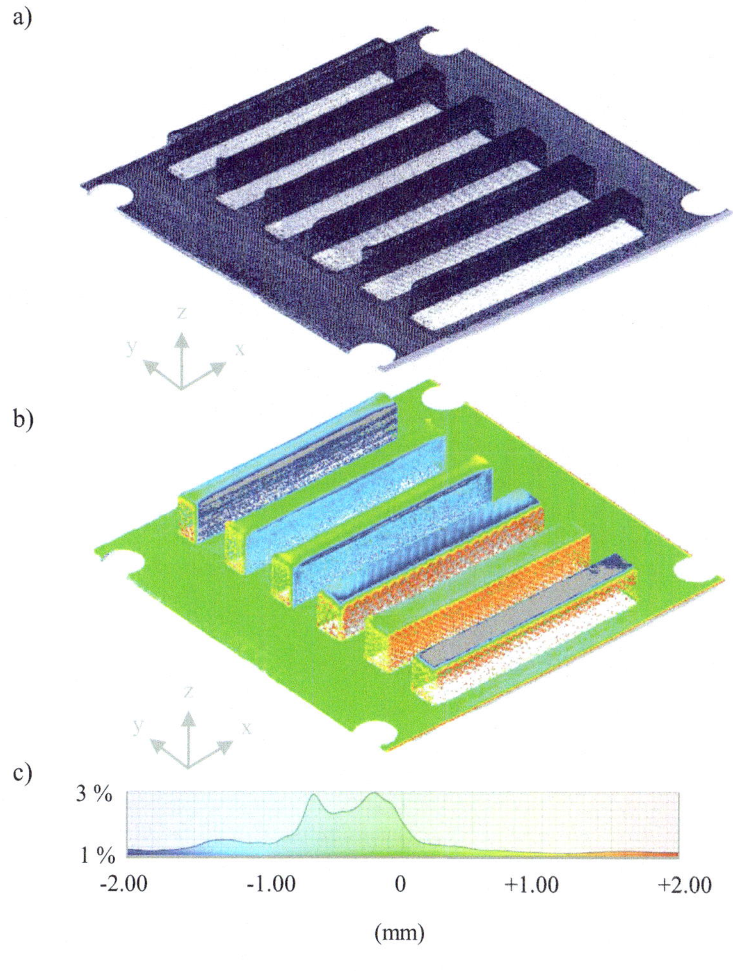

Figure 6.16 a) Scanned point cloud of the built specimens (heat treated) with base plate using CMM b) Measured geometrical deviation between point cloud and CAD geometry c) Histogram of the measured deviation

The geometrical characterization was carried out using a coordinate measuring machine (CMM). Wenzel Präzision GmbH's mid-sized LH87 was attached with a Renishaw PH10M motorized probe head and a 3D shape tracer (laser triangulation sensor). Several scan paths were programmed to ensure collection of data points from all sides. The scanning process then generated a point cloud represented by x, y and z coordinates at a rate of 48,000 points per second with an accuracy of 20 µm. The generated point cloud was then compared with the CAD model using Point Master 5.0 software, resulting in a 3D geometrical deviation plot. Figure 6.16 shows the point cloud generated using the laser triangulation sensor (a), 3D geometrical deviation plot (b) and the distribution of the measured deviation as a histogram (c). From the histogram, a process- and toolpath strategy-independent manufacturing tolerance range of the laser deposition process can be deduced. The tolerance, however, is very much dependent on the toolpath creates, which in turn is dependent on slicing and hatch parameters.

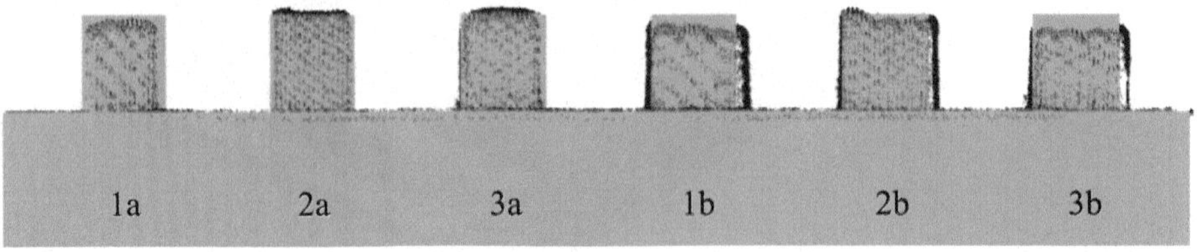

Figure 6.17 Cross-section view of the generated point cloud overlapped with CAD geometry

Figure 6.17 provides usable information on the interaction between toolpath strategy and the slicing and hatch parameters, especially the slicing thickness and path overhang that affects the total height and width of the build geometry. Depending on the tolerance achieved, an additional toolpath can be either added or removed. From Figure 6.17 it can be deduced that for parallel toolpath strategy (1a) an additional toolpath is required and for zig-zag toolpaths (1b and 3b) additional layers are required to achieved the desired geometry.

6.2.4. Mechanical Characterization

Static mechanical characterization of the test geometry includes porosity measurement, hardness measurement and tensile testing. For porosity and hardness measurement, an additional set of three wall geometries for the three process parameter sets were deposited with parallel toolpath strategy. These geometries were deposited on a 250 x 120 x 20 mm^3 substrate, followed by stress-relief heat treatment and sand blasting. The wall geometry was then sectioned along the cutting plane as shown in Figure 6.20, embedded and prepared for micrograph analysis. The cross-section micrographs (un-etched) of the test geometries are shown in Figure 6.18.

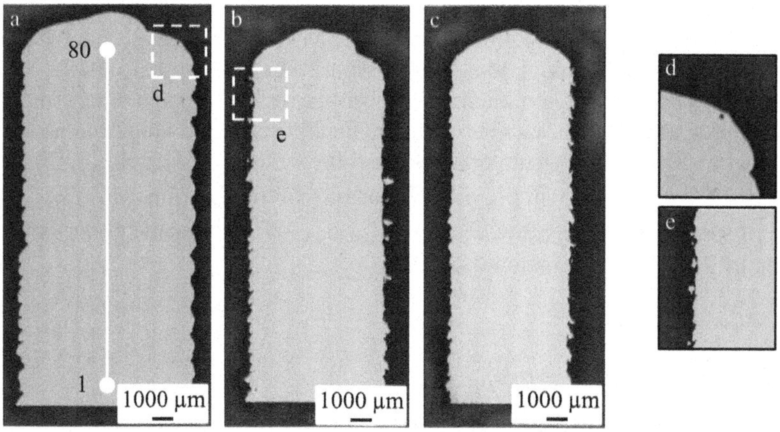

Figure 6.18 Un-etched cross-section micrographs of wall geometry with process parameter: set 1 (a), set 2 (b) and set 3 (c).

6.2.4.1. Porosity

Voids such as pores, cracks and binding defects can reduce the structural properties of a component in terms of both static and dynamic strength. The formation of voids or pores could be due to one of the following mechanisms that takes place during the deposition process.

- The dissolved gases in the molten phase gives rise to pore formation, as the high solidification rate prevents complete outgassing. These pores are typically spherical in shape (see Figure 6.19.a). The gases can also be introduced into the melt pool by remnant atmospheric components in the process zone [156, 157].
- Pores can also be created by powder particles that are not completely melted thereby resulting in interparticle voids [158, 159] (see Figure 6.19.b). The appearance of this porosity is often irregular. This is caused by an insufficient match between the powder properties and the process parameters that doesn't ensure sufficient energy input for melting [160].
- Irregular and large voids can be caused from insufficient hatch overlap or certain toolpath strategies. These are often recursive in nature, aspherical in shape and can reduce the static and dynamic properties of the components to a large extent (see Figure 6.19.c). A systematic identification of process parameters, slicing and hatch parameters coupled with optimized toolpaths can reduce or eliminate these voids.
- Entrapped gas inside a single powder particle that results from uncontrolled or poor particle atomization process (see Figure 6.19.d).

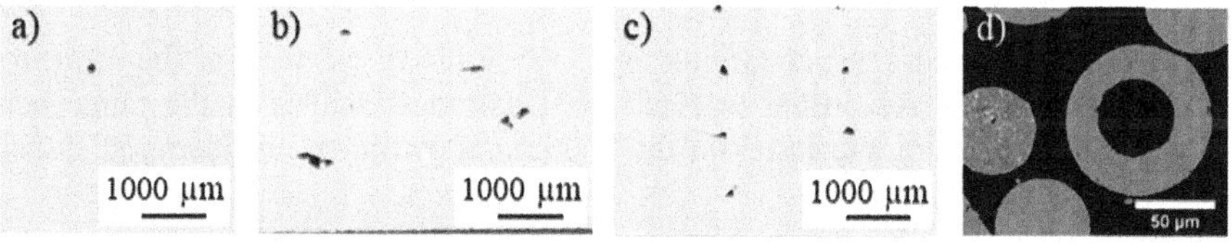

Figure 6.19 Typical voids that generate during the deposition process: a) gas pores b) unmelt powder particles c) poor bonding effect d) entrapped gas in powder particle

In the following section, the micrographs are investigated for the occurring porosity shape as well as the quantity of porosity. Porosity has been measured from the micrograph of unetched specimens through image processing. For this purpose, a light microscopic measurement of the density on micrographs was carried out. Image processing is used to adjust the grey scale values in order to automatically segregate voids and pores from the solid material. The resulting sum of areas of all the voids and pores A_{pores} divided by the total inspected area A_{total} will provide the porosity of the specimen, $\varphi_{specimen}$. From the measured specimen porosity, the density, $\rho_{specimen}$ of the laser metal deposited component can be derived (see equation 6.2).

$$\varphi_{specimen} = \frac{\sum A_{pores}}{A_{total}} \tag{6.1}$$

$$\rho_{specimen} = 1 - \varphi_{specimen} \tag{6.2}$$

Measured porosity values and the resulting density values are shown in Figure 6.20 along with the cross-section micrographs of the etched specimens.

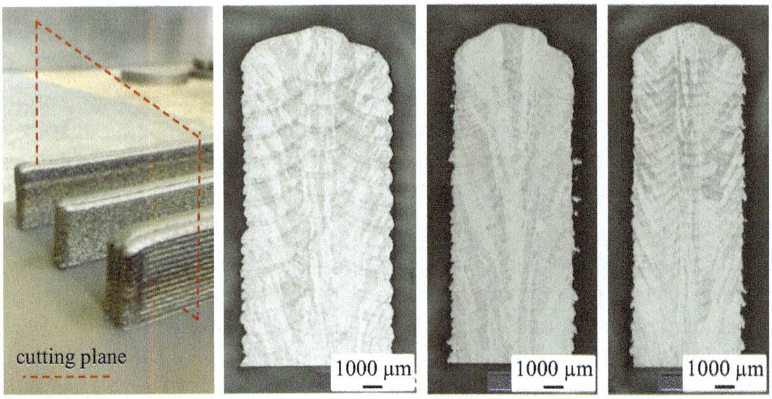

Process parameter	Set 1	Set 2	Set 3
Porosity (%)	0.022	0.012	0.009
Density (%)	99.978	99.988	99.991
Hardness (HV10)	355	356	356

Figure 6.20 Cross-sectional micrographs of wall geometry with parallel toolpaths for three different process parameter sets used for porosity and hardness measurement

6.2.4.2. Vickers hardness

The Vickers hardness test according to ISO6507 [161] was used as the measurement method in this work. According to [162, 163], hardness values in the range between 300 HV and 400 HV are witnessed for the titanium alloy under consideration, depending on the microstructure present.

To measure the hardness profile, the cross-section micrographs used for porosity measurement was employed. Vickers hardness (HV10) was measured at 80 measurement points along the build direction as marked in Figure 6.18.a. For this purpose, an automated insertion of test indentations with a test load of HV10 was performed on the Struers DuraScan 70 hardness testing machine. The system is equipped with a light microscope to measure and record the indentation along with an image evaluation system for automatic measurement of the Vickers hardness. Figure 6.21 plots the HV10 hardness measured along the build direction for the three wall geometries. No specific pattern has been recorded along the build direction within a test geometry or between the specimens. However, the hardness value of the wall geometry built with process parameter set 1 shows a higher statistical range. The average hardness (arithmetic mean) for the test geometries is shown in Figure 6.20. No significant difference have been noticed among the three process parameter sets.

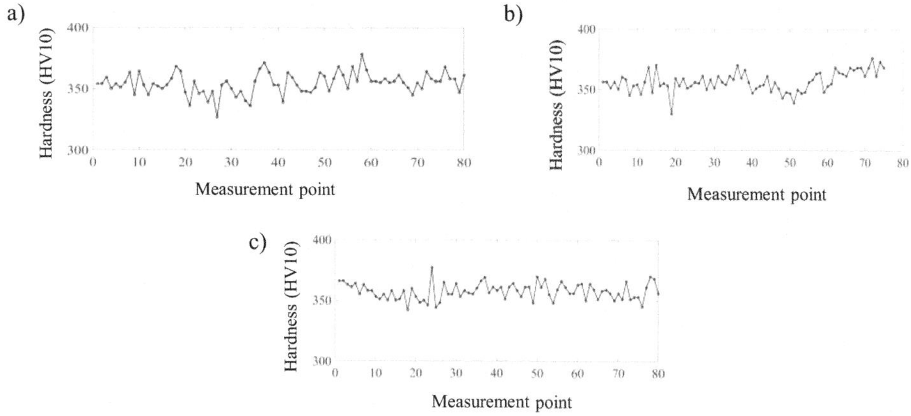

Figure 6.21 Hardness profile along the build direction for test geometry built with process parameter: a) set 1, b) set 2 and c) set 3

6.2.4.3. Static tensile test

Static tensile test is a basic destructive mechanical test used to quantify the material's resistance to tensile forces until fracture. For mechanical characterization, tensile test is essential as it measures modulus of elasticity, ultimate tensile strength, yield strength and the elongation at break. For the tests, 4 Type E flat tensile specimens with 3 mm cross-section thickness were machined from the deposited blocks according to DIN50125 [164], after heat treatment. The tensile tests were carried out using a Zwick Z100 materials testing machine in accordance with ISO6892 [165]. The flat specimens were clamped in the tensile testing machine and the extensometers are attached to the specimen to record the strains.

Figure 6.22 and Figure 6.23 shows the results of the static tensile test carried out for the test geometries. Test specimens fabricated with process parameter set 3 show a higher value of elongation at break in comparison to other two process parameter sets. Also, zig-zag toolpaths show comparatively poor tensile properties with respect to parallel toolpath strategy. This shows clearly that the chosen toolpath strategy and the process parameter set defines the resulting mechanical properties.

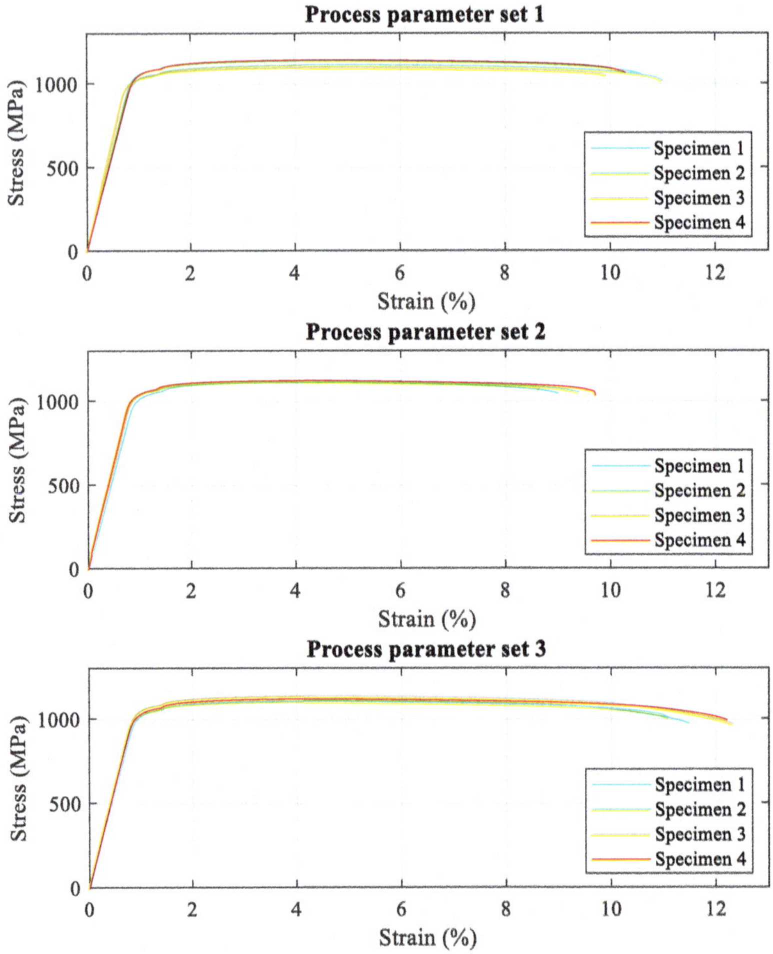

Figure 6.22 Measured stress-strain curve of the test specimens fabricated with parallel toolpath strategy and different process parameter sets

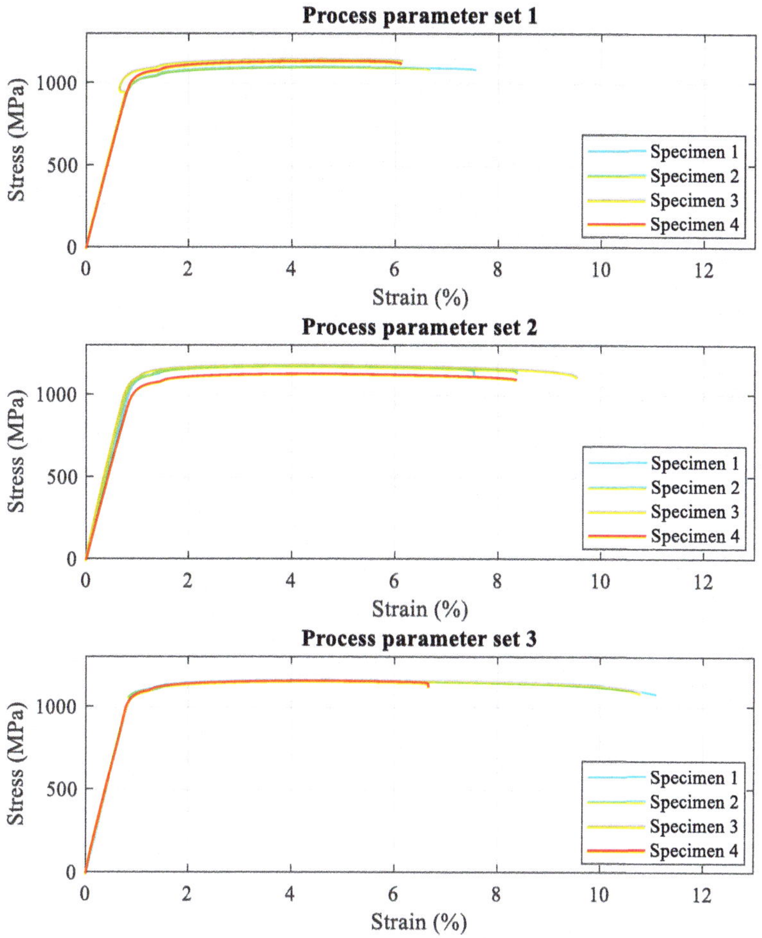

Figure 6.23 Measured stress-strain curve of the test specimens fabricated with zig-zag toolpath strategy and different process parameter sets

6.2.5. Deposition Rate and Productivity Analysis

During the fabrication of the test specimens described in section 6.2.2, parameters including deposition time, non-weld time and total powder utilized were measured. After the deposition process, deposited volume has been calculated through CMM along with the derivation of the deposited weight. Based on these values, the deposition rate, laser utilization rate and the powder utilization rate has been calculated and tabulated in Table 6.5.

Table 6.5 Analysis of deposition rate and productivity based on the fabrications of test specimens

Parameter	**Unit**	**Process parameter Set 1**		**Process parameter Set 1**		**Process parameter Set 3**	
Toolpath strategy	-	Parallel	Zig-zag	Parallel	Zig-zag	Parallel	Zig-zag
Total production time	s	1410	1740	5040	5100	1380	1500
Powder flowrate	g/min	13.1	13.1	6	6	16.1	16.1
Total powder sprayed	g	307.85	379.9	504	510	370.3	402.5
Number of layers	-	15	13	36	29	29	24
Wait time per layer	s	0	0	0	0	0	0
Powder start & end time	s	25	25	25	25	25	25
Transfer/ rapid time	s	28.12	6.5	67.5	14.5	54.37	12
Non-welding time	s	53.12	31.5	92.-5	39.5	79.37	37
Deposition time	s	1356.87	1708.5	4947.5	5060.5	1300.62	1463
Laser utilization rate	%	96.23	98.19	98.16	99.23	94.25	97.53
Usable powder	g	296.25	373.02	494.75	506.05	349.00	392.57
Deposited volume (measured)	mm^3	45647.55	59266.2	49369.5	59682	51100.5	54076.5
Deposited volume (measured)	cm^3	45.64	59.26	49.36	59.68	51.10	54.07
Derived weight	g	201.30	261.36	217.71	263.19	225.35	238.47
Deposition rate	cm^3/hr	121.11	124.88	35.92	42.45	141.44	133.06
Powder utilization rate	%	65.39	68.79	43.19	51.60	60.85	59.24

6.3. Fabrication of Thin-Wall Demonstrator Components

The methodology developed and described in section 6.2 has been validated for industrial use-case in this section. Thin-walled aerospace component shown in Figure 5.44.a. has been chosen for fabrication and geometrical characterization of different toolpath strategies described in section 5.5. Primary process parameters identified in section 6.2 has been used for fabrication, followed by geometrical characterization for measurement of manufacturing tolerance necessary for downstream machining process.

6.3.1. Fabrication of different Toolpath Strategies

Figure 6.24 - Figure 6.27 shows the laser metal deposited aerospace demonstrator component with four different toolpath strategies, classified under contour-based and segment-wise.

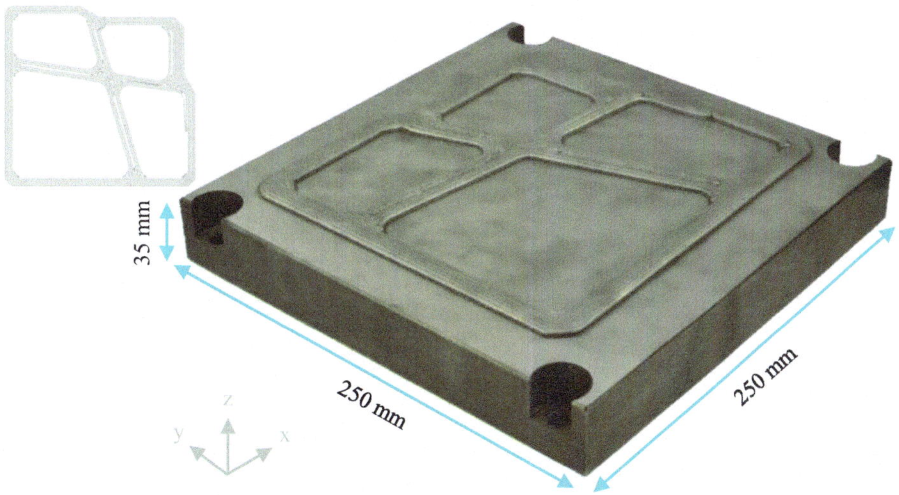

Figure 6.24 Demonstrator component fabricated with contour-based (outside-in) toolpath strategy

The outside-in toolpath strategy was fabricated only for the first three layers due to the fact that the build height deviation within a single layer was highly sufficient enough to cause process instability (Figure 6.24). Moreover, as expected, several voids were visible right after the fabrication of first layer. Rest of the other three strategies were fabricated completely.

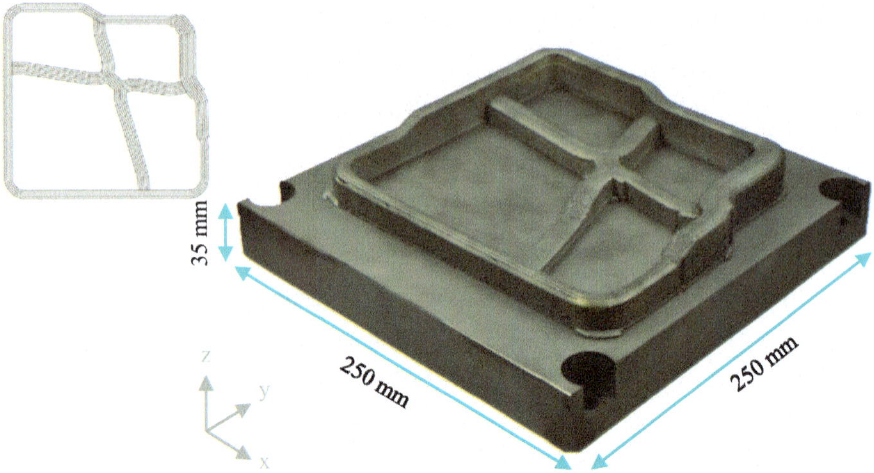

Figure 6.25 Demonstrator component fabricated with contour-based (inside-out) toolpath strategy

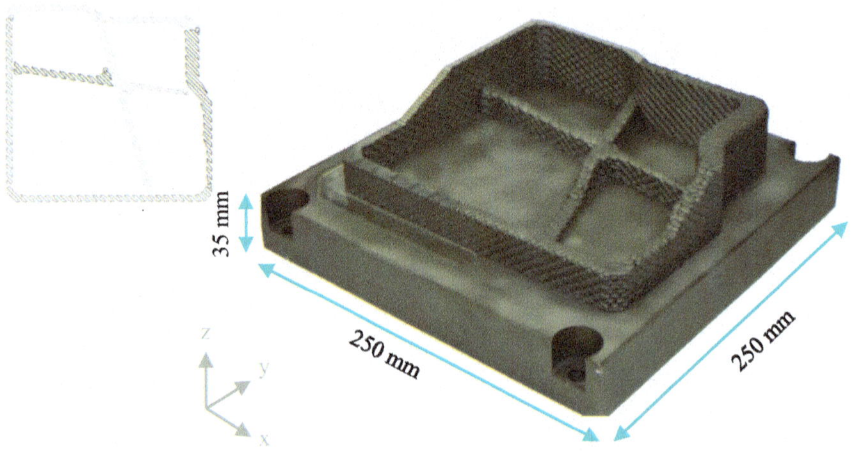

Figure 6.26 Demonstrator component fabricated with segment-wise (wall-wise) toolpath strategy

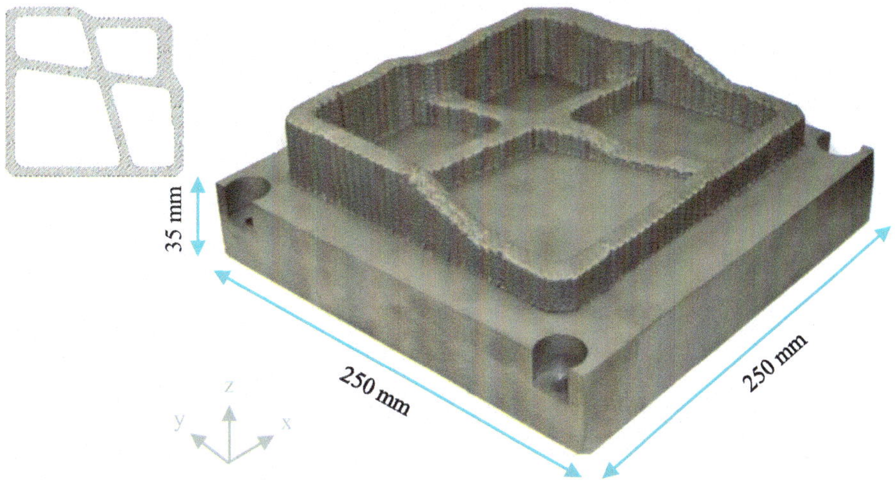

Figure 6.27 Demonstrator component fabricated with segment-wise (feature-wise) toolpath strategy

Visual observations during and after the process showed that the selected toolpath strategy have an influence on the process stability and on the build resolution. Demonstrators built with zig-zag toolpaths have near perfect or sharp corners with equal build height distribution within a layer and throughout the build height (Figure 6.26 and Figure 6.27). However, long parallel toolpaths tend to round-up at the corners and edges with decreasing build height around the edges (outer toolpaths) over the build height (Figure 6.25). This phenomenon was observed during the fabrication of test blocks (compare with Figure 6.17).

Figure 6.28 compares qualitatively the four different toolpath strategies localized to three distinct design features of the thin-wall geometry. It can be noticed that two toolpath strategies, namely contour-based inside-out and segment-wise feature-based toolpaths have produced better fabrication results (numbered 2 and 4 in Figure 6.28). It can be concluded that the contour based outside-in toolpath (numbered 1) is not suitable for thin-wall geometries due to the large voids seen between the toolpaths. This qualitative analysis based on visual inspection is not a sufficient measure to completely access the different toolpath strategies. The fabricated geometries must be subjected to quantitative geometrical characterization, which is described in the following section. Additionally, a non-destructive computed tomography scan to reveal the inner defects and voids is necessary, which however, is beyond the scope of this research work.

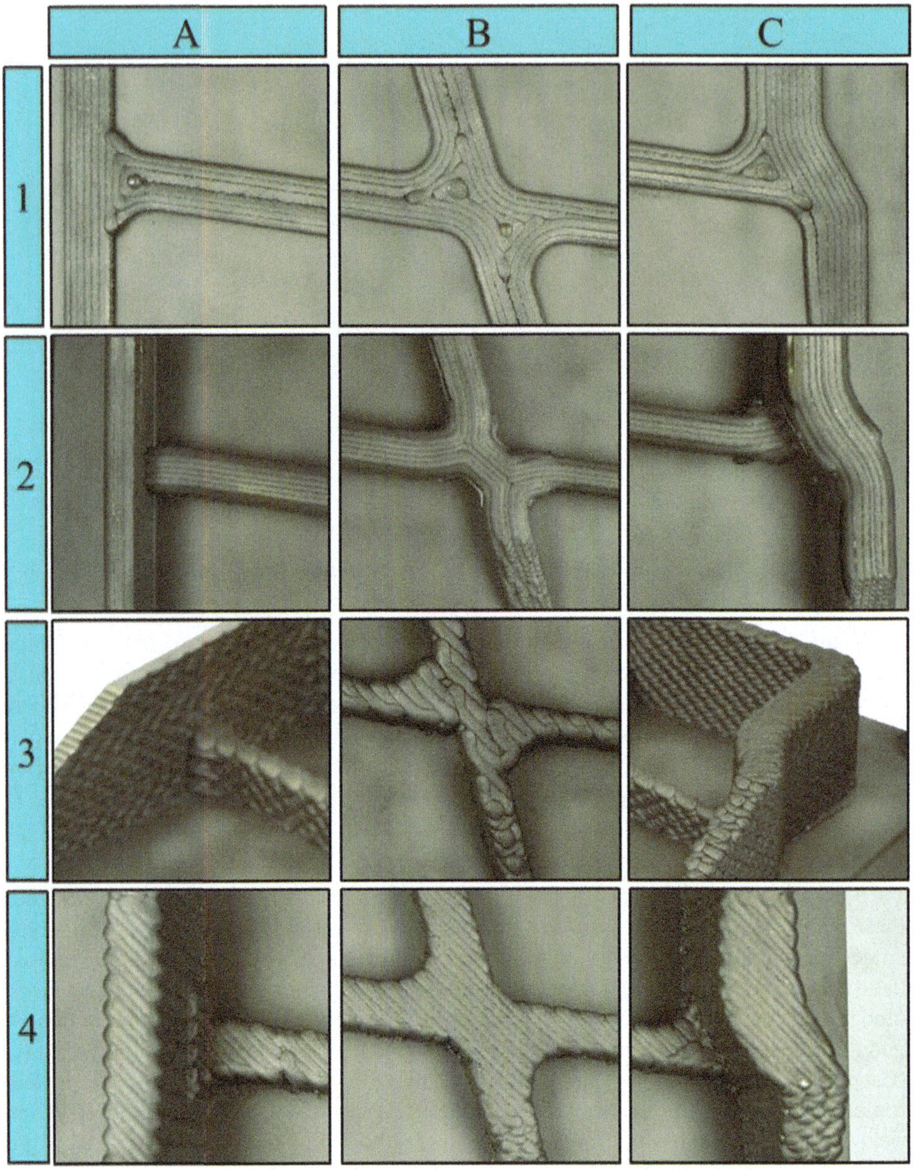

Figure 6.28 Comparison of fabrication results localized at three specific locations of the demonstrator between the different toolpath strategies

6.3.2. Geometrical Characterization

The geometrical characterization was once again carried out using Wenzel Präzision GmbH's mid-sized LH87 coordinate measuring machine (CMM) as described in section 6.2.3. The generated point cloud was compared with the CAD model using Point Master 5.0 software, resulting in a 3D geometrical deviation plot as depicted in the following figures.

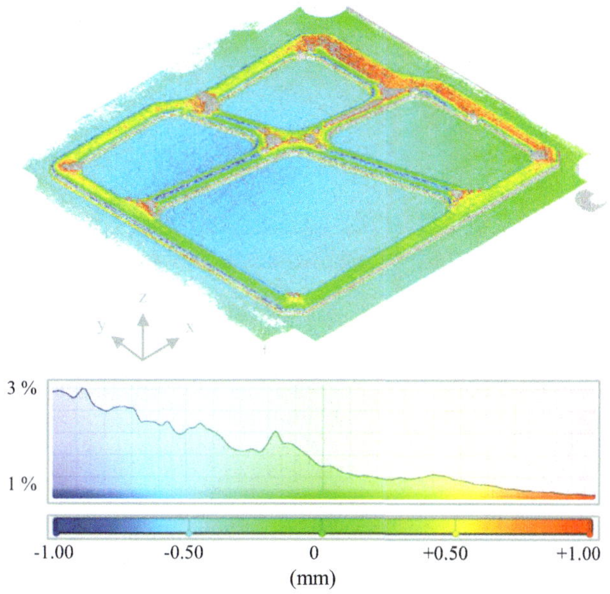

Figure 6.29 Geometrical deviation of the demonstrator geometry shown in Figure 6.24 measured using coordinate measuring machine

Regions of higher build areas and voids are quantified in the CMM plot and the resulting histogram (see Figure 6.29). The shift of the histogram curve towards the left of the graphs depicts that the voids are more prominent than that of the peaks.

In Figure 6.30, the rounding effect of the parallel toolpaths over the build height along the edges are quantified along with regions of insufficient material deposition.

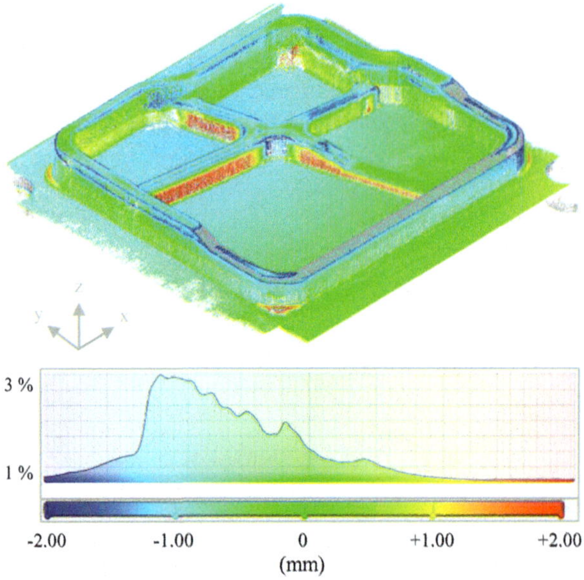

Figure 6.30 Geometrical deviation of the demonstrator geometry shown in Figure 6.25 measured using coordinate measuring machine

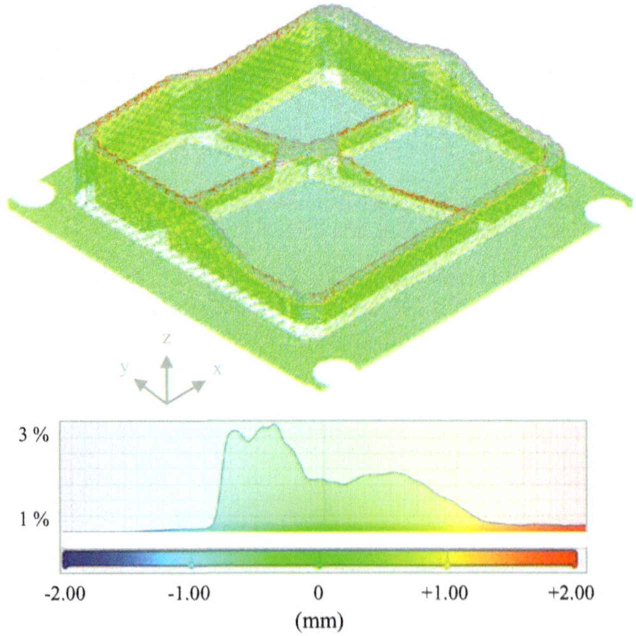

Figure 6.31 Geometrical deviation of the demonstrator geometry shown in Figure 6.26 measured using coordinate measuring machine

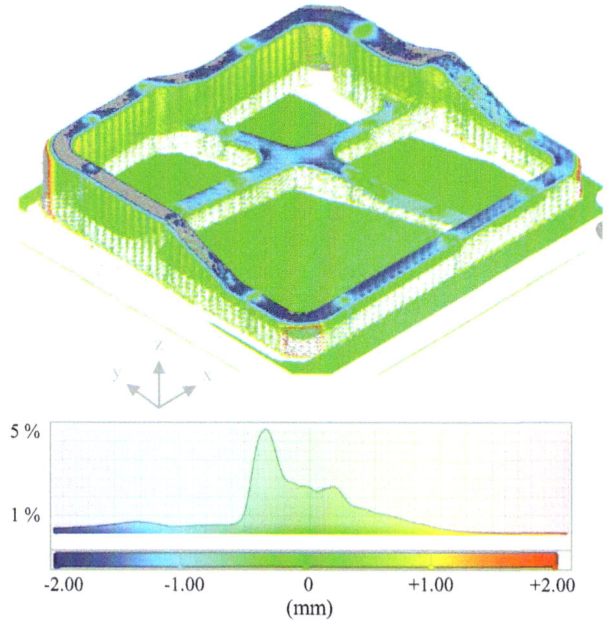

Figure 6.32 Geometrical deviation of the demonstrator geometry shown in Figure 6.27 measured using coordinate measuring machine

The histogram of the demonstrator built with segment-wise feature-based zig-zag toolpath strategy has the minimum deviation from the planned geometry. The overlap of segments is visible with relatively high material deposition areas. However, this particular toolpath strategy provides the least manufacturing tolerance. As already mentioned, a non-destructive computed tomography scan of the intersection and the overlap area will be necessary to access the inner geometrical defects.

Based on the results, a manufacturing tolerance of ± 1.50 mm has been achieved for the chosen demonstrator component independent of the toolpath strategy and process parameter set. A toolpath-strategy specific systematically optimized build strategy can bring down the manufacturing tolerance to ± 0.75 mm. This fulfills the set goal of ± 1.0 mm manufacturing tolerance or the geometrical accuracy of the as-built geometry.

7. Conclusion and Outlook

7.1. Conclusion

Prior to the time- and cost-intensive certification and qualification steps necessary for the serial production of aerospace DED components, a stable and repeatable process is required. In addition, a CAD/CAM system that is capable of generating toolpaths specific to process, deposition rate and geometry is inevitable for achieving defect-free components with high deposition rate DED processes. However, the existing commercial software tools do not satisfy the varying requirements of the various DED processes. Additionally, the existing tools for toolpath planning and optimization require human intervention.

This research work, therefore, has developed six different geometry-specific toolpath generation algorithms focusing on automation of CAD data pre-processing, including volume decomposition, segmentation and feature-specific toolpath generation. The developed automation algorithms have reduced the build job preparation time from hours to minutes. The six different component clusters considered in this work include bulk volume, multi-directional, gradually inclined, thin-wall, non-planar parametric and non-planar freeform. These six clusters encompass the majority of the DED geometries.

Usage of STEP-file format in contrast to the STL-file format has provided advantages in terms of higher-order representation of geometries suitable for efficient pre-processing, especially for non-planar and freeform geometries. Also, STEP-file format supported NURBS provide a more accurate representation of the geometry in comparison to the tessellated geometry. This allows for generation of more accurate toolpaths. However, it has been noticed that the final geometrical accuracy of the additively manufactured part depends not only on the data format, but also on the process parameter definition, toolpath strategy, deposition rate and the controller of the kinematic system used. The Open Cascade Technology based python framework developed in this thesis combines the advantages of using the geometry kernel's existing classes and methods along with the ease of implementation of new algorithms and functions using python programming language.

Of the several intrinsic and extrinsic process parameters that influence the build quality of the DED component, the influences of primary process parameters namely laser power, deposition velocity and powder flow rate on the resulting weld bead geometry have been studied systematically. The results of which were used for determination of toolpath planning parameters (i.e. slicing and hatching parameters). Furthermore, the influence of several toolpath strategies in thin-wall geometries have been analyzed in detail. For one chosen thin-wall demonstrator geometry, four different toolpath strategies were implemented followed by its fabrication through powder-based laser metal deposition process. The fabricated demonstrators were then subjected to quantitative and qualitative geometrical characterization, as well as mechanical characterization.

V. J. Prakash, *Automated Process Planning for Additive Manufacturing with Directed Energy Deposition*, Light Engineering für die Praxis,
https://doi.org/10.1007/978-3-031-87666-0_7

7.2. Outlook

The several algorithms and functions developed in this research work have been implemented and tested within a python framework. As an outlook of this research work, these algorithms can be implemented into existing commercial CAD/CAM software packages mentioned in section 2.5.2., thereby extending their capability.

On the other hand, the in-house process planning platform mentioned in section 5.6 can be extended further with more functionalities and a better GUI. The current version relies on human input to choose the right algorithm for the geometry to be sliced, i.e. component clustering is not automated which can be developed further. Furthermore, an interface and interaction with an external kinematic simulation model and a thermal simulation model can be developed to iteratively optimize toolpaths for reachability, equal heat distribution and minimum residual stresses.

Toolpath generation and its automation algorithms developed throughout this research work are based on computational-geometry methods. However, one recent research work has used artificial intelligence to compute toolpaths especially for thin-wall geometries [122]. A Reinforcement learning framework has been used to determine boundary boxes and to generate toolpaths after parameterization (pixelation) of the layer data. Though the developed methodology has been proven suitable only for thin-wall geometries fabricated with single track, the potential of using AI techniques has been demonstrated. Usage of such AI methods and techniques especially for optimization of toolpath for temperature distribution could potentially replace expensive multi-physics simulation tools.

The ability to work with STEP-file format-based CAD models, the flexibility and the open architecture of the developed python framework could be extended to other AM technologies like powder bed fusion. Transfer of know-how to build-data preparation in powder-bed processes, especially the developed NURBS-based toolpath generation algorithms can enable geometry-dependent laser vector generation that can result in additive manufacturing of components with higher geometrical accuracy.

In order to access the print quality of the different toolpath strategies developed, physical fabrication tests are still required, causing machine downtime. A CAD/CAM based dexel- or voxel-based volumetric simulation model can complement the process planning platform by evaluating the various toolpath strategies prior to physical fabrication.

References

1. Chi K (2015) A look at Global Metallic Fastener Development in Two Airline Manufacturing Giants, Boing and Airbus. Industry Focus 2015:178–182

2. Huang R, Riddle M, Graziano D, Warren J, Das S, Nimbalkar S, Cresko J, Masanet E (2016) Energy and emissions saving potential of additive manufacturing: the case of lightweight aircraft components. Journal of Cleaner Production 135:1559–1570. https://doi.org/10.1016/j.jclepro.2015.04.109

3. Emmelmann C, Sander P, Kranz J, Wycisk E (2011) Laser Additive Manufacturing and Bionics: Redefining Lightweight Design. Physics Procedia 12:364–368. https://doi.org/10.1016/j.phpro.2011.03.046

4. Dutta B, Froes FH (2017) The Additive Manufacturing (AM) of titanium alloys. Metal Powder Report 72:96–106. https://doi.org/10.1016/j.mprp.2016.12.062

5. Gralow M, Weigand F, Herzog D, Wischerhopp T, Emmelmann C (2020) Biomimetic design and laser additive manufacturing—A perfect symbiosis? Journal of Laser Applications 32:21201. https://doi.org/10.2351/1.5131642

6. Gibson I, Rosen D, Stucker B (2015) Additive Manufacturing Technologies. Springer New York. https://doi.org/10.1007/978-1-4939-2113-3

7. Lindemann C, Koch R (2016) Cost Efficient Design and Planning for Additive Manufacturing Technologies. In: 27th Annual International Solid FreeForm Fabrication Symposium

8. Gebhardt A (2016) Additive Fertigungsverfahren: Additive Manufacturing und 3D-Drucken für Prototyping - Tooling - Produktion. Hanser; Ciando, München

9. Gebhardt A (2013) Generative Fertigungsverfahren: Additive Manufacturing und 3D Drucken für Prototyping - Tooling - Produktion. Hanser; Ciando, München

10. Associates W (2022) Wohlers report 2022: 3D printing and additive manufacturing global state of the industry. Wohlers Associates, Fort Collins (Colo.)

11. Candel-Ruiz A, Kaufmann S, Müllerschön O (2015) Strategies for high deposition rate additive manufacturing by Laser Metal Deposition. In: Lasers in Manufacturing Conference 2015

V. J. Prakash, *Automated Process Planning for Additive Manufacturing with Directed Energy Deposition*, Light Engineering für die Praxis,
https://doi.org/10.1007/978-3-031-87666-0

12. Sciaky, Inc. EBAM Drives Innovation for Many Applications and Industries. https://www.sciaky.com/additive-manufacturing/applications-industries

13. Piscopo G, Iuliano L (2022) Current research and industrial application of laser powder directed energy deposition. The International Journal of Advanced Manufacturing Technology. https://doi.org/10.1007/s00170-021-08596-w

14. Möller MLB (2021) Prozessmanagement für das Laser-Pulver-Auftragschweißen. Springer Berlin, Heidelberg. https://doi.org/10.1007/978-3-662-62225-4

15. Duarte VR (2022) Developments in Directed Energy Deposition Additive Manufacturing: In-situ Hot Forging and Indirect Tooling. Dissertation, Nova School of Science and Technology

16. Romanenko D, Prakash VJ, Kuhn T, Moeller C, Hintze W, Emmelmann C (2022) Effect of DED process parameters on distortion and residual stress state of additively manufactured Ti-6Al-4V components during machining. Procedia CIRP 111:271–276. https://doi.org/10.1016/j.procir.2022.08.020

17. Norsk Titanium AS (2023) Norsk Titanium: Company Presentation. https://www.norsktitanium.com/storage/home/20230328-Norsk-Titanium-Corporate-Presentation_vE.pdf

18. Munsch M, Schmidt-Lehr M, Wycisk E, Führer T (2022) AMPOWER Report 2022: Additive manufacturing Market Report. AMPOWER GmbH & Co. KG

19. Dordlofva C, Lindwall A, Törlind P (2016) Opportunities and Challenges for Additive Manufacturing in Space Applications. In: NordDesign 2016, vol 6

20. DebRoy T, Mukherjee T, Wei HL, Elmer JW, Milewski JO (2021) Metallurgy, mechanistic models and machine learning in metal printing. Nature Reviews Materials 6:48–68. https://doi.org/10.1038/s41578-020-00236-1

21. Ahn D-G (2021) Directed Energy Deposition (DED) Process: State of the Art. International Journal of Precision Engineering and Manufacturing-Green Tech 8:703–742. https://doi.org/10.1007/s40684-020-00302-7

22. Norsk Titanium AS (2017) Norsk Titanium Delivers First FAA-Certified, Additive Manufactured Ti64 Structural Aviation Components https://www.norsktitanium.com/media/press/norsk-titanium-delivers-first-faa-certified-additive-manufactured-ti64-structural-aviation-components

23. DIN EN ISO/ASTM 52900:2022-03, Additive Fertigung- Grundlagen-Terminologie (ISO/ASTM 52900:2021)

24. ASTM F3187-16. Standard Guide for Directed Energy Deposition of Metals. ASTM International

25. Buchfink G (2007) The laser as a tool: A light beam conquers industrial production. Vogel Buchverlag, Würzburg, Germany

26. Thompson SM, Bian L, Shamsaei N, Yadollahi A (2015) An overview of Direct Laser Deposition for additive manufacturing; Part I: Transport phenomena, modeling and diagnostics. Additive Manufacturing 8:36–62. https://doi.org/10.1016/j.addma.2015.07.001

27. Saboori A, Gallo D, Biamino S, Fino P, Lombardi M (2017) An Overview of Additive Manufacturing of Titanium Components by Directed Energy Deposition: Microstructure and Mechanical Properties. Applied Sciences 7:883. https://doi.org/10.3390/app7090883

28. Dass A, Moridi A (2019) State of the Art in Directed Energy Deposition: From Additive Manufacturing to Materials Design. Coatings 9:418. https://doi.org/10.3390/coatings9070418

29. Chakraborty D, Tirumala T, Chitral S, Sahoo BN, Kiran DV, Kumar PA (2022) The State of the Art for Wire Arc Additive Manufacturing Process of Titanium Alloys for Aerospace Applications. Journal of Material Engineering and Performance 31:6149–6182. https://doi.org/10.1007/s11665-022-07128-1

30. Li Z, Sui S, Ma X, Tan H, Zhong C, Bi G, Clare A, Gasser A, Chen J (2022) High deposition rate powder- and wire-based laser directed energy deposition of metallic materials: A review. International Journal of Machine Tools and Manufacture 181:103942. https://doi.org/10.1016/j.ijmachtools.2022.103942

31. Digital Alloys (2019) Directed Energy Deposition (DED). https://www.digitalalloys.com/blog/directed-energy-deposition/

32. Lia F, Park J, Tressler J, Martukanitz R (2017) Partitioning of laser energy during directed energy deposition. Additive Manufacturing 18:31–39. https://doi.org/10.1016/j.addma.2017.08.012

33. Weisheit A, Gasser A, Backes G, Jambor T, Pirch N, Wissenbach K (2013) Direct Laser Cladding , Current Status and Future Scope of Application. In: Majumdar

JD, Manna I (eds) Laser-Assisted Fabrication of Materials, vol 161. Springer Berlin, Heidelberg, pp 221–240. https://doi.org/10.1007/978-3-642-28359-8_5

34. Toyserkani E, Khajepour A, Corbin SF (2004) Laser Cladding. CRC Press. https://doi.org/10.1201/9781420039177

35. Heigel JC (2018) Thermo-Mechanical Modeling of Thin Wall Builds using Powder Fed Directed Energy Deposition. In: Thermo-Mechanical Modeling of Additive Manufacturing. Elsevier, pp 137–15. https://doi.org/10.1016/B978-0-12-811820-7.00010-0

36. Nabhani M, Razavi RS, Barekat M (2018) An empirical-statistical model for laser cladding of Ti-6Al-4V powder on Ti-6Al-4V substrate. Optics & Laser Technology 100:265–271. https://doi.org/10.1016/j.optlastec.2017.10.015

37. Wolff SJ, Lin S, Faierson EJ, Liu WK, Wagner GJ, Cao J (2017) A framework to link localized cooling and properties of directed energy deposition (DED)-processed Ti-6Al-4V. Acta Materialia 132:106–117. https://doi.org/10.1016/j.actamat.2017.04.027

38. Stavropoulos P, Foteinopoulos P (2018) Modelling of additive manufacturing processes: a review and classification. Manufacturing Review 5:2. https://doi.org/10.1051/mfreview/2017014

39. Abbes B, Anedaf T, Abbes F, Li Y (2021) Direct energy deposition metamodeling using a meshless method. Engineering Computations 38:1226–1240. https://doi.org/10.1108/EC-10-2019-0447

40. Hayes BJ, Martin BW, Welk B, Kuhr SJ, Ales TK, Brice DA, Ghamarian I, Baker AH, Haden CV, Harlow DG, Fraser HL, Collins PC (2017) Predicting tensile properties of Ti-6Al-4V produced via directed energy deposition. Acta Materialia 133:120–133. https://doi.org/10.1016/j.actamat.2017.05.025

41. Wei HL, Mukherjee T, Zhang W, Zuback JS, Knapp GL, De A, DebRoy T (2021) Mechanistic models for additive manufacturing of metallic components. Progress in Materials Science 116:100703. https://doi.org/10.1016/j.pmatsci.2020.100703

42. Zhao Y, Jia Y, Chen S, Shujun S, Shi J, Li F (2020) Process planning strategy for wire-arc additive manufacturing: Thermal behavior considerations. Additive Manufacturing 32:100935. https://doi.org/10.1016/j.addma.2019.100935

43. Lepski D, Brückner F (2009) Laser Cladding. In: Dowden J.(ed) The Theory of Laser Materials Processing, vol 119. Springer Netherlands, Dordrecht, pp 235–279. https://doi.org/10.1007/978-1-4020-9340-1_8

44. Ye J, Bab-Hadiashar A, Hoseinnezhad R, Alam N, Vargas-Uscategui A, Patel M, Cole I (2022) Predictions of in-situ melt pool geometric signatures via machine learning techniques for laser metal deposition. International Journal of Computer Integrated Manufacturing:1–17. https://doi.org/10.1080/0951192X.2022.2048422

45. Wang X, Jiang J, Xia C, Yu Y (2022) Macroscopic and Microstructural Features of Metal Thin-Wall Fabricated by Laser Material Deposition: A Review. Coatings 12:1104. https://doi.org/10.3390/coatings1208110

46. Pinkerton AJ, Li L (2004) Modelling the geometry of a moving laser melt pool and deposition track via energy and mass balances. Journal of Physics D: Applied Physics 37:1885–1895. https://doi.org/10.1088/0022-3727/37/14/003

47. El Cheikh H, Courant B, Hascoët J-Y, Guillén R (2012) Prediction and analytical description of the single laser track geometry in direct laser fabrication from process parameters and energy balance reasoning. Journal of Materials Processing Technology 212:1832–1839. https://doi.org/10.1016/j.jmatprotec.2012.03.016

48. Tabernero I, Lamikiz A, Ukar E, Martínez S (2012) Geometry Modelling of Clads Generated by Laser Cladding. Material Science Forum 713:85–90. https://doi.org/10.4028/www.scientific.net/MSF.713.85

49. Mazumder J (2015) Design for Metallic Additive Manufacturing Machine with Capability for "Certify as You Build". Procedia CIRP 36:187–192. https://doi.org/10.1016/j.procir.2015.01.009

50. Zeinali M, Khajepour A (2010) Height Control in Laser Cladding Using Adaptive Sliding Mode Technique: Theory and Experiment. Journal of Manufacturing Science and Engineering 132. https://doi.org/10.1115/1.4002023

51. Tang L, Landers RG (2011) Layer-to-Layer Height Control for Laser Metal Deposition Process. Journal of Manufacturing Science and Engineering 133. https://doi.org/10.1115/1.4003691

52. Amini M, Chang SI (2018) MLCPM: A process monitoring framework for 3D metal printing in industrial scale. Computers & Industrial Engineering 124:322–330. https://doi.org/10.1016/j.cie.2018.07.041

53. Chabot A, Rauch M, Hascoët J-Y (2019) Towards a multi-sensor monitoring methodology for AM metallic processes. Weld World 63:759–769. https://doi.org/10.1007/s40194-019-00705-4

54. Baier D, Wolf F, Weckenmann T, Lehmann M, Zaeh MF (2022) Thermal process monitoring and control for a near-net-shape Wire and Arc Additive Manufacturing. Production Engineering 16:811–822. https://doi.org/10.1007/s11740-022-01138-7

55. Hauser T, Reisch RT, Kamps T, Kaplan A, Volpp J (2022) Acoustic emissions in directed energy deposition processes. The International Journal of Advanced Manufacturing Technology. https://doi.org/10.1007/s00170-021-08598-8

56. Smoqi Z, Bevans BD, Gaikwad A, Craig J, Abul-Haj A, Roeder B, Macy B, Shield JE, Rao P (2022) Closed-loop control of meltpool temperature in directed energy deposition. Materials & Design 215:110508. https://doi.org/10.1016/j.matdes.2022.110508

57. Jeon I, Sohn H (2022) Online melt pool depth estimation in laser metal deposition using a coaxial thermography system. Journal of Laser Applications 34:22001. https://doi.org/10.2351/7.0000618

58. Singh P, Dutta D (2001) Multi-Direction Slicing for Layered Manufacturing. Computer-Aided Design 1:129. https://doi.org/10.1115/1.1375816

59. Ruan J, Sparks TE, Panackal A, Liou FW, Eiamsa-ard K, Slattery K, Chou H-N, Kinsella M (2007) Automated Slicing for a Multiaxis Metal Deposition System. Journal of Manufacturing Science and Engineering 129:303. https://doi.org/10.1115/1.2673492

60. Liou F, Ruan J, E Sparks T (2010) Multi-Aautoaxis Planning System (MAPS) for Hybrid Laser Metal Deposition Processes. 21st Annual International Solid Freeform Fabrication Symposium - An Additive Manufacturing Conference, SFF 2010

61. Yang Y, Fuh JYH, Loh HT, Wong YS (2003) A Volumetric Difference-based Adaptive Slicing and Deposition Method for Layered Manufacturing. Journal of Manufacturing Science and Engineering 125:586–594. https://doi.org/10.1115/1.1581887

62. Yang Y, Fuh J, Loh HT, Wong YS (2003) Multi-orientational deposition to minimize support in the layered manufacturing process. Journal of Manufacturing Systems 22:116–129. https://doi.org/10.1016/S0278-6125(03)90009-4

63. Lee K, Jee H (2015) Slicing algorithms for multi-axis 3-D metal printing of overhangs. Journal of Mechanical Science and Technology 29:5139–5144. https://doi.org/10.1007/s12206-015-1113-y

64. Dolenc A, Mäkelä I (1994) Slicing procedures for layered manufacturing techniques. Computer-Aided Design 26:119–126. https://doi.org/10.1016/0010-4485(94)90032-9

65. Ding D, Pan Z, Cuiuri D, Li H, Larkin N, van Duin S (2015) Multi-direction slicing of STL models for robotic wire-feed additive manufacturing. In: 26th Annual International Solid FreeForm Fabrication Symposium

66. Ding D, Pan Z, Cuiuri D, Li H, Larkin N, van Duin S (2016) Automatic multi-direction slicing algorithms for wire based additive manufacturing. Robotics and Computer-Integrated Manufacturing 37:139–150. https://doi.org/10.1016/j.rcim.2015.09.002

67. Muraleedharan LP, Kannan SS, Karve A, Muthuganapathy R (2018) Random cutting plane approach for identifying volumetric features in a CAD mesh model. Computers & Graphics 70:51–61. https://doi.org/10.1016/j.cag.2017.07.025

68. Guo H, Xu J, Zhang S, Zhang Y, Tan J (2023) Multi-orientation optimization of complex parts based on model segmentation in additive manufacturing. Journal of Mechanical Science and Technology 37:317–331. https://doi.org/10.1007/s12206-022-1231-2

69. Xiao X, Xiao H (2021) Autonomous Robotic Feature-Based Freeform Fabrication Approach. Materials (Basel) 15. https://doi.org/10.3390/ma15010247

70. Xiangping W, Haiou Z, Guilan W, Lingpeng, W (2014) Adaptive slicing for multi-axis hybrid plasma deposition and milling. In: 25th Annual International Solid FreeForm Fabrication Symposium

71. Chalvin M, Campocasso S, Baizeau T, Hugel V (2019) Automatic multi-axis path planning for thinwall tubing through robotized wire deposition. Procedia CIRP 79:89–94. https://doi.org/10.1016/j.procir.2019.02.017

72. Makem JE, Fogg HJ, Mukherjee N (2020) Automatic Feature Recognition Using the Medial Axis for Structured Meshing of Automotive Body Panels. Computer-Aided Design 124:102845. https://doi.org/10.1016/j.cad.2020.102845

73. Zhao G, Ma G, Xiao W, Tian Y (2019) Feature-based five-axis path planning method for robotic additive manufacturing. Proceedings of the Institution of

Mechanical Engineers, Part B: Journal of Engineering Manufacture 233:1412–1424. https://doi.org/10.1177/0954405417752508

74. Zhao G, Ma G, Feng J, Xiao W (2018) Nonplanar slicing and path generation methods for robotic additive manufacturing. The International Journal of Advanced Manufacturing Technology 96:3149–3159. https://doi.org/10.1007/s00170-018-1772-9

75. Tata K, Fadel G, Bagchi A, Aziz N (1998) Efficient slicing for layered manufacturing. Rapid Prototyping Journal 4:151–167. https://doi.org/10.1108/13552549810239003

76. Choi SH, Kwok KT (2002) A tolerant slicing algorithm for layered manufacturing. Rapid Prototyping Journal 8:161–179. https://doi.org/10.1108/13552540210430997

77. Kirschman C, Jara-Almonte CC (1992) A Parallel Slicing Algorithm for Solid Freeform Fabrication Processes. In: International Solid Freeform Fabrication Symposium, pp 26–33

78. Chalasani KL, Grogan B, Bagchi A, Jara-Almonte CC, Ogale AA, Dooley RL (1991) An Algorithm to Slice 3-D Shapes for Reconstruction in Prototyping Systems. In: ASME 1991 Computers in Engineering Conference. American Society of Mechanical Engineers, pp 209–215. https://doi.org/10.1115/CIE1991-0027

79. Luo RC, Chang YC, Tzou JH (2001) The development of a new adaptive slicing algorithm for layered manufacturing system. In: International Conference on Robotics and Automation, pp 1334–1339. https://doi.org/10.1109/ROBOT.2001.932795

80. Flores J, Garmendia I, Pujana J (2019) Toolpath generation for the manufacture of metallic components by means of the laser metal deposition technique. International Journal of Advanced Manufacturing Technology 101:2111–2120. https://doi.org/10.1007/s00170-018-3124-1

81. Sundaram R, Choi J (2001) A slicing procedure for 5-axis layered manufacturing. International Solid Freeform Fabrication Symposium. https://doi.org/10.26153/tsw/3334

82. Ruan J, Tang L, Liou FW, Landers RG (2010) Direct Three-Dimensional Layer Metal Deposition. Computer-Aided Design 132:64502. https://doi.org/10.1115/1.4002624

83. Jin YA, He Y, Fu JZ (2013) An Adaptive Tool Path Generation for Fused Deposition Modeling. Advanced Materials Research 819:7–12. https://doi.org/10.4028/www.scientific.net/AMR.819.7

84. Zhang J (2004) Adaptive Slicing for a Multi-Axis Laser Aided Manufacturing Process. Computer-Aided Design 126:254. https://doi.org/10.1115/1.1649966#

85. Kerschbaumer M, Ernst G (2004) Hybrid manufacturing process for rapid high performance tooling combining high speed milling and laser cladding. In: International Congress on Applications of Lasers & Electro-Optics. Laser Institute of America, p 1710. https://doi.org/10.2351/1.5060234

86. Jee H, Kim M (2017) Spherically curved layer (SCL) model for metal 3-D printing of overhangs. Journal of Mechanical Science and Technology. 31:5729–5735. https://doi.org/10.1007/s12206-017-1114-0

87. Starly B, Lau A, Sun W, Lau W, Bradbury T (2005) Direct slicing of STEP based NURBS models for layered manufacturing. Computer-Aided Design 37:387–397

88. Zhou MY (2005) STEP-based approach for direct slicing of CAD models for layered manufacturing. International Journal of Production Research 43:3273–3285. https://doi.org/10.1080/00207540500097809

89. Jamieson R, Hacker H (1995) Direct slicing of CAD models for rapid prototyping. Rapid Prototyping Journal 1:4–12. https://doi.org/10.1108/13552549510086826

90. Shi Y, Chen X, Cai D, Huang S (2004) Application software system based on direct slicing for rapid prototyping. International Journal of Production Research 42:2227–2242. https://doi.org/10.1080/00207540410001671688

91. Fadel G, Kirschman C (1994) A NURBS based approach for rapid product realization. Fifth International Conference on Rapid Prototyping:229–240

92. T.Hayasi M, Asiabanpour B (2009) Machine path generation using direct slicing from design-by-feature solid model for rapid prototyping. The International Journal of Advanced Manufacturing Technology 45. https://doi.org/10.1007/s00170-009-1944-8

93. Gohari H, Barari A, Kishawy H (2018) An efficient methodology for slicing NURBS surfaces using multi-step methods. The International Journal of Advanced Manufacturing Technology 95:3111–3125. https://doi.org/10.1007/s00170-017-1219-8

94. Park SC, Choi BK (2000) Tool-path planning for direction-parallel area milling. Computer-Aided Design 32:17–25. https://doi.org/10.1016/S0010-4485(99)00080-9

95. Jin Y, He Y, Xue G, Fu J (2015) A parallel-based path generation method for fused deposition modeling. The International Journal of Advanced Manufacturing Technology 77:927–937. https://doi.org/10.1007/s00170-014-6530-z

96. Wang X, Wang A, Li Y (2019) A sequential path-planning methodology for wire and arc additive manufacturing based on a water-pouring rule. The International Journal of Advanced Manufacturing Technology 103:3813–3830. https://doi.org/10.1007/s00170-019-03706-1

97. Faust RC, Minetto R, Volpato N (2023) Parallel tool-path generation for Additive Manufacturing: A GPU-based zigzag filling. Advances in Industrial and Manufacturing Engineering 6:100107. https://doi.org/10.1016/j.aime.2022.100107

98. Yang Y, Loh HT, Fuh J, Wang YG (2002) Equidistant path generation for improving scanning efficiency in layered manufacturing. Rapid Prototyping Journal 8:30–37. https://doi.org/10.1108/13552540210413284

99. Jin GQ, Li WD, Gao L (2013) An adaptive process planning approach of rapid prototyping and manufacturing. Robotics and Computer-Integrated Manufacturing 29:23–38. https://doi.org/10.1016/j.rcim.2012.07.001

100. Ganesan M, Fadel GM (1994) Hollowing rapid prototyping parts using offsetting techniques. Proceedings of the fifth international conference on rapid prototyping

101. Farouki RT, Koenig T, Tarabanis KA, Korein JU, Batchelder JS (1995) Path planning with offset curves for layered fabrication processes. Journal of Manufacturing Systems 14:355–368. https://doi.org/10.1016/0278-6125(95)98872-4

102. Papacharalampopoulos A, Bikas H, Stavropoulos P (2018) Path planning for the infill of 3D printed parts utilizing Hilbert curves. Procedia Manufacturing 21:757–764. https://doi.org/10.1016/j.promfg.2018.02.181

103. Palfrader P, Held M (2015) Computing Mitered Offset Curves Based on Straight Skeletons. Computer-Aided Design and Applications 12:414–424. https://doi.org/10.1080/16864360.2014.997637

104. Ding D, Pan Z, Cuiuri D, Li H (2015) A practical path planning methodology for wire and arc additive manufacturing of thin-walled structures. Robotics and

Computer-Integrated Manufacturing 34:8–19. https://doi.org/10.1016/j.rcim.2015.01.003

105. Ding Y, Akbari M, Kovacevic R (2018) Process planning for laser wire-feed metal additive manufacturing system. The International Journal of Advanced Manufacturing Technology 95:355–365. https://doi.org/10.1007/s00170-017-1179-z

106. Messing A, Roschli A, K. Post B, J.Love L (2017) Using skeletons for void filling in large-scale additive manufacturing. In: 28th Annual International Solid FreeForm Fabrication Symposium

107. Jee S, Koo T (2003) Tool-path generation for NURBS surface machining. In: Proceedings of the 2003 American Control Conference, 2003. IEEE, pp 2614–2619. https://doi.org/10.1109/ACC.2003.1243471

108. Lasemi A, Xue D, Gu P (2010) Recent development in CNC machining of freeform surfaces: A state-of-the-art review. Computer-Aided Design 42:641–654. https://doi.org/10.1016/j.cad.2010.04.002

109. Zhao G, Zavalnyi O, Liu Y, Xiao W (2018) Prospects for using T-splines for the development of the STEP-NC-based manufacturing of freeform surfaces. The International Journal of Advanced Manufacturing Technology 43:3703. https://doi.org/10.1007/s00170-018-3182-4

110. Feng H-Y, Li H (2002) Constant scallop-height tool path generation for three-axis sculptured surface machining. Computer-Aided Design 34:647–654. https://doi.org/10.1016/S0010-4485(01)00136-1

111. Tournier C, Lartigue C (2008) 5-axis Iso-scallop Tool Paths along Parallel Planes. Computer-Aided Design and Applications 5:278–286. https://doi.org/10.3722/cadaps.2008.278-286

112. Can A, Ünüvar A (2010) A novel iso-scallop tool-path generation for efficient five-axis machining of free-form surfaces. The International Journal of Advanced Manufacturing Technology 51:1083–1098. https://doi.org/10.1007/s00170-010-2698-z

113. Sarkar S, Dey PP (2014) A new iso-parametric machining algorithm for free-form surface. Proceedings of the Institution of Mechanical Engineers, Part E: Journal of Process Mechanical Engineering 228:197–209. https://doi.org/10.1177/0954408913495191

114. Liu W, Zhu S-M, Huang T, Zhou C (2020) An efficient iso-scallop tool path generation method for three-axis scattered point cloud machining. The International Journal of Advanced Manufacturing Technology 107:3471–3483. https://doi.org/10.1007/s00170-020-05188-y

115. Jin Y, He Y, Fu J, Gan W, Lin Z (2014) Optimization of tool-path generation for material extrusion-based additive manufacturing technology. Additive Manufacturing 1-4:32–47. https://doi.org/10.1016/j.addma.2014.08.004

116. Li R, Zhang H, Dai F, Huang C, Wang G (2019) End lateral extension path strategy for intersection in wire and arc additive manufactured 2319 aluminum alloy. Rapid Prototyping Journal 26:360–369. https://doi.org/10.1108/RPJ-05-2019-0123

117. Ma G, Zhao G, Li Z, Xiao W (2019) A Path Planning Method for Robotic Wire and Arc Additive Manufacturing of Thin-Walled Structures with Varying Thickness. IOP Conference Series: Materials Science and Engineering 470:12018. https://doi.org/10.1088/1757-899X/470/1/012018

118. Michel F, Lockett H, Ding J, Martina F, Marinelli G, Williams S (2019) A modular path planning solution for Wire + Arc Additive Manufacturing. Robotics and Computer-Integrated Manufacturing 60:1–11. https://doi.org/10.1016/j.rcim.2019.05.009

119. Nguyen L, Buhl J, Bambach M (2020) Continuous Eulerian tool path strategies for wire-arc additive manufacturing of rib-web structures with machine-learning-based adaptive void filling. Additive Manufacturing 35:101265. https://doi.org/10.1016/j.addma.2020.101265

120. Rauch M, Hascoet J-Y, Querard V (2021) A Multiaxis Tool Path Generation Approach for Thin Wall Structures Made with WAAM. Journal of Manufacturing and Materials Processing 5:128. https://doi.org/10.3390/jmmp5040128

121. Li C, Wu B, Zhang Z, Zhang Y (2023) A novel process planning method of 3 + 2-axis additive manufacturing for aero-engine blade based on machine learning. Journal of Intelligent Manufacturing 34:2027–2042. https://doi.org/10.1007/s10845-021-01898-6

122. Petrik J, Bambach M (2023) Reinforcement learning and optimization based path planning for thin-walled structures in wire arc additive manufacturing. Journal of Manufacturing Processes 93:75–89. https://doi.org/10.1016/j.jmapro.2023.03.013

123. Murua M, Suarez A, Galar D, Santana R (2020) Tool-Path Problem in Direct Energy Deposition Metal-Additive Manufacturing: Sequence Strategy Generation. IEEE Access 8:91574–91585. https://doi.org/10.1109/ACCESS.2020.2994748

124. Xiong Y, van Campen A, van Vlierberghe A, Kempen K, Kruth JP (2017) Time-optimal scan path planning based on analysis of sliced geometry. In: 28th Annual International Solid FreeForm Fabrication Symposium

125. Spranger F, Graf B, Schuch M, Hilgenberg K, Rethmeier M (2018) Build-up strategies for additive manufacturing of three dimensional Ti-6Al-4V-parts produced by laser metal deposition. Journal of Laser Applications 30:22001. https://doi.org/10.2351/1.4997852

126. Gardner JA, Nethercott-Garabet T, Kaill N, Campbell RI, Bingham AG, Engstrom DS, Balc NO (2019) Aligning Material Extrusion Direction with Mechanical Stress via 5-Axis Tool Paths. In: International Solid Freeform Symphosium

127. Zhang H, Yao Y, Ma Y, Lackner M, Jiang Y (2022) A 3D printing tool-path generation strategy based on the partition of principal stress field for fused filament fabrication. The International Journal of Advanced Manufacturing Technology 122:1719–1735. https://doi.org/10.1007/s00170-022-09957-9

128. (2018) Grundlegende wissenschaftliche Konzepterstellung zu bestehenden Herausforderungen und Perspektiven für die additive Fertigung mit Lichtbogen: Studie im Auftrag der Forschungsvereinigung Schweißen und verwandte Verfahren e.V. des DVS. DVS-Berichte, Band 345. DVS Media, Düsseldorf

129. Munsch M, Schmidt-Lehr M, Wycisk E, Führer T (2020) Additive Manufacturing: New Metal Technologies

130. Heilemann M, Lingner M (2022) Hybrid Additive manufaturing: Fraunhofer Research-Institution for Additive Manufacturing Technology IAPT. Additive Alliance 2022 Study Report

131. Zhang C, Shen C, Hua X, Li F, Zhang Y, Zhu Y(2020) Influence of wire-arc additive manufacturing path planning strategy on the residual stress status in one single buildup layer. The International Journal of Advanced Manufacturing Technology 111:797–806. https://doi.org/10.1007/s00170-020-06178-w

132. Song G-H, Lee C-M, Kim D-H (2021) Investigation of Path Planning to Reduce Height Errors of Intersection Parts in Wire-Arc Additive Manufacturing. Materials (Basel) 14. https://doi.org/10.3390/ma14216477

133. Mao Y, Wei H, Chang L, Yue J, Liu T, Xing F, Liao W (2023) Origin of deposition errors and layer-wise control strategies during laser additive manufacturing. Virtual and Physical Prototyping 18. https://doi.org/10.1080/17452759.2023.2173615

134. Sun X, Mazur M, Cheng C-T (2023) A review of void reduction strategies in material extrusion-based additive manufacturing. Additive Manufacturing 67:103463. https://doi.org/10.1016/j.addma.2023.103463

135. Gibson BT, Mhatre P, Borish MC, Atkins CE, Potter JT, Vaughan JE, Love LJ (2022) Controls and process planning strategies for 5-axis laser directed energy deposition of Ti-6Al-4V using an 8-axis industrial robot and rotary motion. Additive Manufacturing 58:103048. https://doi.org/10.1016/j.addma.2022.103048

136. Lu X, Cervera M, Chiumenti M, Lin X (2021) Residual Stresses Control in Additive Manufacturing. Journal of Manufacturing and Materials Processing 5:138. https://doi.org/10.3390/jmmp5040138

137. Odermatt AE, Dorn F, Ventzke V (2022) Coaxial laser directed energy deposition with wire of thin-walled duplex stainless steel parts: Process discontinuities and their impact on the mechanical properties. CIRP Journal of Manufacturing Science and Technology 37:443–453. https://doi.org/10.1016/j.cirpj.2022.02.019

138. Zhang W, Abbott WM, Sasnauskas A, Lupoi R (2022) Process Parameters Optimisation for Mitigating Residual Stress in Dual-Laser Beam Powder Bed Fusion Additive Manufacturing. Metals 12:420. https://doi.org/10.3390/met12030420

139. Weisz-Patrault D, Margerit P, Constantinescu A (2022) Residual stresses in thin walled-structures manufactured by directed energy deposition: in-situ measurements, fast thermo-mechanical simulation and buckling. Additive Manufacturing 56:102903. https://doi.org/10.1016/j.addma.2022.102903

140. Derekar KS, Ahmad B, Zhang X, Joshi SS, Lawrence J, Xu L, Melton G, Addison A (2022) Effects of Process Variants on Residual Stresses in Wire Arc Additive Manufacturing of Aluminum Alloy 5183. Journal of Manufacturing Science and Engineering 144. https://doi.org/10.1115/1.4052930

141. Bremer J, Walderich P, Pirch, Schleifenbaum JH, Gasser A, Schopphoven T (2021) Effects of path accuracy on additively manufactured specimens by laser material deposition using six-axis robots. Journal of Laser Applications 33:12045. https://doi.org/10.2351/7.0000308

142. Bingol OR, Krishnamurthy A (2019) NURBS-Python: An open-source object-oriented NURBS modeling framework in Python. SoftwareX 9:85–94. https://doi.org/10.1016/j.softx.2018.12.005

143. Open Cascade Technology Documentation. https://dev.opencascade.org/

144. Piegl LA, Tiller W (1997) The NURBS book, 2nd ed. Monographs in visual communication. Springer, New York

145. Lim S, Buswell RA, Valentine PJ, Piker D, Austin SA, Kestelier X (2016) Modelling curved-layered printing paths for fabricating large-scale construction components. Additive Manufacturing 12:216–230. https://doi.org/10.1016/j.addma.2016.06.004

146. Can A, Ünüvar A (2010) A novel iso-scallop tool-path generation for efficient five-axis machining of free-form surfaces. The International Journal of Advanced Manufacturing Technology 51:1083–1098. https://doi.org/10.1007/s00170-010-2698-z

147. Ding D, Pan Z, Cuiuri D, Li H, van Duin S (2016) Advanced Design for Additive Manufacturing: 3D Slicing and 2D Path Planning. In: Shishkovsky IV (ed) New Trends in 3D Printing

148. Chang Y-C, Kao J-H, Pinilla JM, Dong J, Prinz FB (2001) Medial Axis Transform (MAT) of General 2D Shapes and 3D Polyhedra for Engineering Applications. In: Kimura F (ed) Geometric Modelling. Springer US, Boston, MA, pp 37–52. https://doi.org/10.1007/978-0-387-35490-3_3

149. Petrat T, Winterkorn R, Graf B, Gumenyuk A, Rethmeier M (2018) Build-up strategies for temperature control using laser metal deposition for additive manufacturing. Weld World 62:1073–1081. https://doi.org/10.1007/s40194-018-0604-8

150. Biegler M, Wang J, Kaiser L, Rethmeier M (2020) Automated Tool-Path Generation for Rapid Manufacturing of Additive Manufacturing Directed Energy Deposition Geometries. steel research int 91:2000017. https://doi.org/10.1002/srin.202000017

151. Köhler M, Sun L, Hensel J, Pallaspuro S, Kömi J, Dilger K, Zhang Z (2021) Comparative study of deposition patterns for DED-Arc additive manufacturing of Al-4046. Materials & Design 210:110122. https://doi.org/10.1016/j.matdes.2021.110122

152. Arana M, Ukar E, Rodriguez I, Aguilar D, Álvarez P (2022) Influence of deposition strategy and heat treatment on mechanical properties and microstructure of 2319 aluminium WAAM components. Materials & Design 221:110974. https://doi.org/10.1016/j.matdes.2022.110974

153. Möller M, Baramsky N, Ewald A, Emmelmann C, Schlattmann J (2016) Evolutionary-based Design and Control of Geometry Aims for AMD-manufacturing of Ti-6Al-4V Parts. Physics Procedia 83:733–742. https://doi.org/10.1016/j.phpro.2016.08.075

154. Prakash VJ, Surrey P, Möller M, Emmelmann C (2018) Influence of adaptive slice thickness and retained heat effect on laser metal deposited thin-walled freeform structures. Procedia CIRP 74:233–237.https://doi.org/10.1016/j.procir.2018.08.101

155. AMS2801C. Heat Treatment of Titanium Alloy Parts. SAE International, 2022

156. Mahamood RM, Akinlabi ET (2015) Modelling of Process Parameters Influence on Degree of Porosity in Laser Metal Deposition Process. In: Transactions on Engineering Technologies. Springer Netherlands, Dordrecht, pp 31–42. https://doi.org/10.1007/978-94-017-9588-3_3

157. Seyda V (2018) Werkstoff- und Prozessverhalten von Metallpulvern in der laseradditiven Fertigung. Springer Berlin Heidelberg, Berlin, Heidelberg. https://doi.org/10.1007/978-3-662-58233-6

158. Sterling AJ, Torries B, Shamsaei N, Thompson SM, Seely DW (2016) Fatigue behavior and failure mechanisms of direct laser deposited Ti–6Al–4V. Materials Science and Engineering: A 655:100–112. https://doi.org/10.1016/j.msea.2015.12.026

159. Wang M, Lin X, Huang W (2015) Laser additive manufacture of titanium alloys. Materials Technology:1–8. https://doi.org/10.1179/1753555715Y.0000000079

160. Kasperovich G, Haubrich J, Gussone J, Requena G (2016) Correlation between porosity and processing parameters in TiAl6V4 produced by selective laser melting. Materials & Design 105:160–170. https://doi.org/10.1016/j.matdes.2016.05.070

161. DIN EN ISO 6507-1:2018-07 Metallische Werkstoffe - Härteprüfung nach Vickers - Teil 1: Prüfverfahren

162. Gerd L, James CW (2007) Titanium. Springer Berlin, Heidelberg. https://doi.org/10.1007/978-3-540-73036-1

163. Peters M (2011) Titan und Titanlegierungen, Wiley-VCH, Weinheim. ISBN- 978-3-527-30539-1

164. DIN 50125:2016-12, Testing of metallic materials - Tensile test pieces

165. ISO 6892-1:2019: Metallic materials - Tensile testing - Part 1: Method of test at room temperature. International Organization for Standardization, Geneva, Switzerland

A. Appendix

A.1. Cost analysis according to part complexity

For two demonstrator geometries varying in their shape complexity, the cost for fabrication (including engineering, pre-processing, material and manufacturing) were enquired for different DED technologies from different suppliers. The table below shows the cost for fabrication along with the normalized cost per cu.cm.

Supplier	DED Technology	Normalized cost per cu.cm, in €	
		Simple	Complex
Supplier 1	Laser metal deposition (powder)	9	20
Supplier 2	Laser metal deposition (wire)	7,25	9,3
Supplier 3	Laser metal deposition (wire)	2,25	2,6
Supplier 4	Wire arc additive manufacturing	3,5	39
Supplier 5	Electron beam additive manufacturing	60	100

V. J. Prakash, *Automated Process Planning for Additive Manufacturing with Directed Energy Deposition*, Light Engineering für die Praxis,
https://doi.org/10.1007/978-3-031-87666-0

A.2. Cross-sectional micrographs

Cross-section micrographs of three single-bead geometries for every process parameter set:

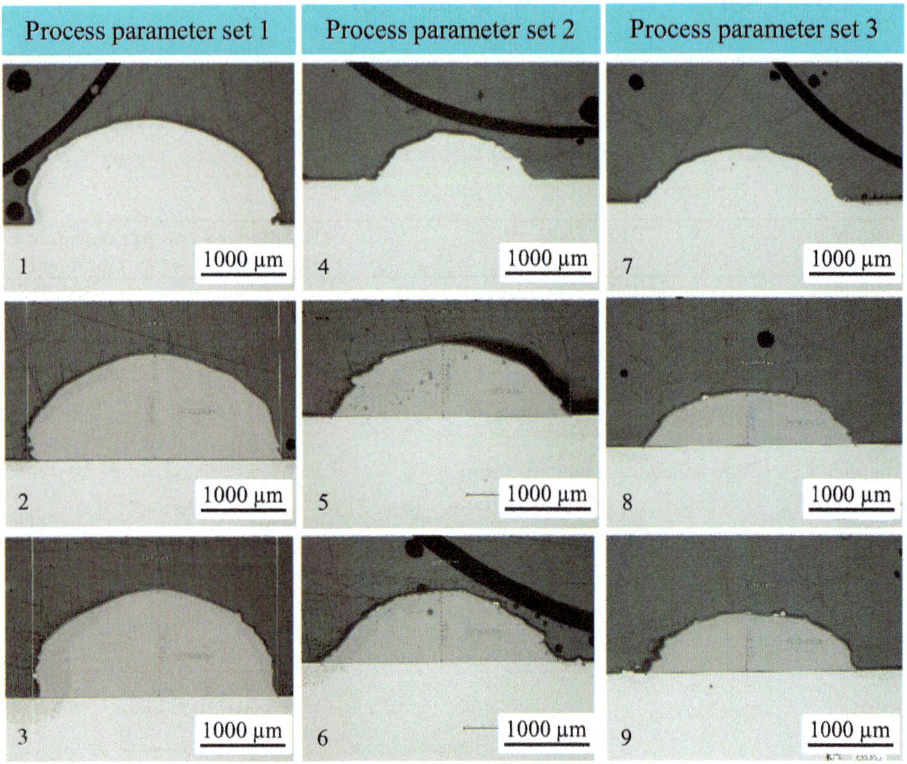

Cross-sectional micrographs of block geometry with parallel toolpaths and 30% overlap for three different process parameter sets along with the measured geometry parameters:

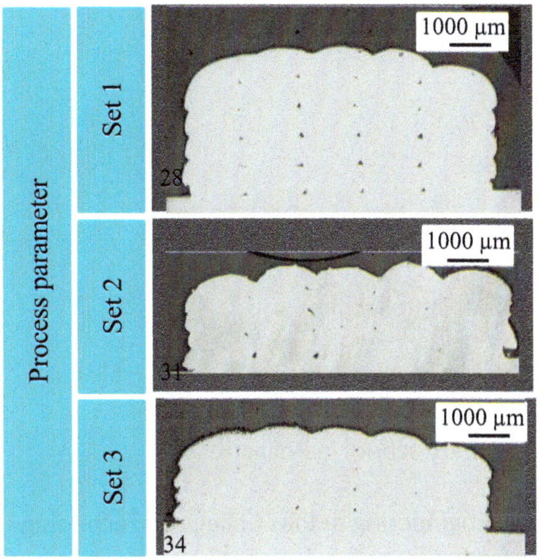

Set 1: $h_5 = 6.17$ mm, $h = 1.23$ mm, $b = 12.86$ mm

Set 2: $h_5 = 2.83$ mm, $h = 0.56$ mm, $b = 9.48$ mm

Set 3: $h_5 = 3.32$ mm, $h = 0.66$ mm, $b = 10.95$ mm

Cross-sectional micrographs of block geometry with parallel toolpaths and 35% overlap for three different process parameter sets along with the measured geometry parameters:

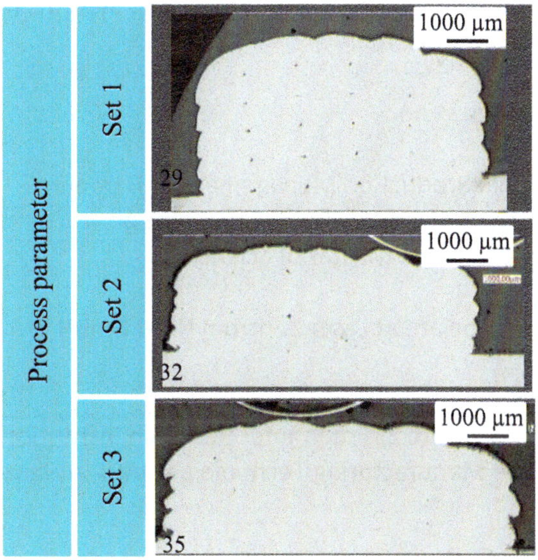

Set 1: $h_5 = 6.53$ mm, $h = 1.30$ mm, $b = 12.81$ mm

Set 2: $h_5 = 2.96$ mm, $h = 0.59$ mm, $b = 8.92$ mm

Set 3: $h_5 = 3.49$ mm, $h = 0.69$ mm, $b = 10.49$ mm

Curriculum Vitae

Surname	Jothi Prakash
First name	Vishnuu
Nationality	Indian
Date of birth	05/12/1991
Place and country of birth	Pollachi, India

06/1997 to 05/2002	Primary school in Pollachi
06/2002 to 05/2007	High school in Pollachi
06/2007 to 06/2009	Higher secondary school in Pollachi
07/2009 to 06/2013	Mechanical Engineering at PSG College of Technology in Coimbatore Qualification: Bachelor of Engineering
07/2013 to 06/2014	Graduate Engineer Trainee at MAHLE Filter Systems (I) Limited in Gurgaon, India
07/2014 to 09/2014	Travel and Vacation
10/2014 to 04/2017	International Production Management at Hamburg University of Technology TUHH in Hamburg Qualification: Master of Science
05/2017 to 12/2017	Research Associate at Laser Zentrum Nord GmbH in Hamburg
01/2018 to present	Research Associate at Fraunhofer Research Institution for Additive Manufacturing Technologies IAPT in Hamburg

V. J. Prakash, *Automated Process Planning for Additive Manufacturing with Directed Energy Deposition*, Light Engineering für die Praxis,
https://doi.org/10.1007/978-3-031-87666-0

The manufacturer's authorised representative in the EU is Springer Nature Customer Service Centre GmbH, Europaplatz 3, 69115 Heidelberg, Germany. If you have any concerns regarding our products, please contact ProductSafety@springernature.com

Printed and bound by CPI Group (UK) Ltd, Croydon, CR0 4YY
01/10/2026
02225946-0001